Elisabetta Princi
Rubber

Also of Interest

Rubber Analysis.
Forrest, 2019
ISBN 978-3-11-064027-4, e-ISBN 978-3-11-064028-1

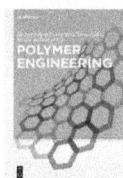

Polymer Engineering
Tylkowski, Wieszczycka, Jastrzab (Eds.), 2017
ISBN 978-3-11-046828-1, e-ISBN 978-3-11-046974-5

Recycling and Re-use of Waste Rubber.
Forrest, 2019
ISBN 978-3-11-064400-5, e-ISBN 978-3-11-064414-2

Chemical Product Technology.
Murzin, 2018
ISBN 978-3-11-047531-9, e-ISBN 978-3-11-047552-4

e-Polymers.
Editor-in-Chief: Seema Agarwal
ISSN 2197-4586
e-ISSN 1618-7229

Elisabetta Princi

Rubber

Science and Technology

DE GRUYTER

Author
Dr Elisabetta Princi
16164 Genova
Italy

ISBN 978-3-11-064031-1
e-ISBN (PDF) 978-3-11-064032-8
e-ISBN (EPUB) 978-3-11-064052-6

Library of Congress Control Number: 2018964956

Bibliographic information published by the Deutsche Nationalbibliothek
The Deutsche Nationalbibliothek lists this publication in the Deutsche Nationalbibliografie;
detailed bibliographic data are available on the Internet at http://dnb.dnb.de.

© 2019 Walter de Gruyter GmbH, Berlin/Boston
Typesetting: Integra Software Services Pvt. Ltd.
Printing and binding: CPI books GmbH, Leck
Cover image: Westend61 / GettyImages

www.degruyter.com

Preface

Currently the use of synthetic rubber is widespread, as the characteristics and properties of these elastomers make them useful in almost all economic sectors: automotive, footwear, civil construction, plastics, medical applications and others that are of crucial importance in the daily life of our society.

The majority of design engineers has little experience with rubber and are sometimes not sure how to select the rubber best suited for their applications, especially since there are so many options. Rubber science and technology comprise both art and science aspects because the base elastomer is highly customisable, together with all the other main ingredients of rubber compound, such as the curing system, fillers, anti-degradants and processing aids among them.

The first step in rubber compound design is the elastomer selection. During this choice, it is important to know to what substances the finished rubber component may come into contact with (oils, ozone, etc.), because some elastomers have properties better suited for certain applications than others. It is also important to understand the end uses and the required physical-mechanical properties. For example, if the end product needs to have high abrasion resistance, there would be specific fillers added into the formulation to help achieve to that property. It is worth to note that there is no rubber compound that will give all of the 'ideal' properties at the same time. One property will have to be compromised in order for another to be made prominent.

The most important part of any formulation is the curing system. There are different types of curing systems, such as sulfuric and peroxide curing, which also influence the final properties of the rubber compound. The smallest differences in the amount or ratio of curing agents can drastically change the properties of the rubber compound. Accelerators are usually part of the curing system as well and are used to shorten the time to cure. Without an accelerator, the rubber compound would cure with the curing ingredients, but it would take exponentially longer time.

The aim of this book is to provide the key information and insights about rubber compounding, transformation technologies, applications and reclaiming methods. Emphasis is given on the scientific approach to rubber science, departing from the description of the basic chemistry and physics of elastomers and elasticity theories, just arriving to a complete overview on the most used rubber materials and thermoplastic elastomers.

https://doi.org/10.1515/9783110640328-201

Contents

1 Basics on rubber

1.1 The rubber history

Long before the arrival of European explorers, rubber was already known to the American native people that were used to produce a number of water-resistant objects using the rubber from the *Hevea brasiliensis* tree [1]. They called this tree 'ca-hu-chu' or 'the crying tree'. Columbus learned during his second voyage to America about a game played by the natives of Haiti in which balls of an elastic 'tree resin' were used. In 1525, Padre d'Anghieria reported that he had seen Mexican tribes playing with elastic balls.

Taking into account the usefulness of the material for wiping lead pencil marks from paper, the first use for rubber in Europe was as eraser, as suggested by Magellan, a descendent of the famous Portuguese navigator. The name 'rubber' was first introduced in England by Joseph Priestley in 1770, who popularised it to the extent that it became known as India Rubber.

Charles de la Condamine approached the first scientific study on rubber when he knew it during his trip to Peru in 1735. Condamine met in Guiana a French engineer named Fresnau who was studying rubber on its home ground. For him rubber was nothing more than a 'type of condensed resinous oil'.

In the nineteenth century a rapid growth in the technical developments and applications of rubber occurred. In 1820, thanks to its waterproofing characteristics, the British industrialist Nadier started the production of rubber threads, attempting to use them in clothing accessories. After few years, in 1823, the British inventor and chemist Macintosh established in Glasgow a plant for manufacturing waterproof clothes and rainproof garments with which his name has become synonymous.

At that time, despite the usefulness of unvulcanised natural rubber for containers, flexible tubing, elastic bands and waterproofing, consumers were discouraged since the cold weather affected such kind of items, leaving them brittle and with a tendency to stick together if left in the sun. Therefore, many attempts to develop a process to upgrade the rubber qualities were carried out, such as including the use of nitric acid, which conversely destroyed the material.

Around 1840 in the United States, Charles Nelson Goodyear accidentally discovered the vulcanisation process founding that natural rubber and sulphur could react together at elevated temperatures leading to a new material. A piece of fabric coated with rubber, sulphur and a few other ingredients was hang close to a stove: the part in direct contact with the stove had greatly changed its properties leading to a surprising result! Almost immediately, Goodyear understood the importance of his finding; thus, he continued the experiments. At that time, he did not know that during vulcanisation the sulphur links attached to the *cis*-1,4-polyisoprene chains of natural rubber form the cross-linked structure (also called network), which is the prerequisite for obtaining the

https://doi.org/10.1515/9783110640328-001

typical elastic properties (Fig. 1.1). Indeed, thanks to this process, rubber acquires increased strength, durability, toughness and elasticity, making it impermeable to gases and resistant to heat, electricity, chemical action and abrasion. Vulcanised rubber also exhibits frictional properties highly desired for tyre applications.

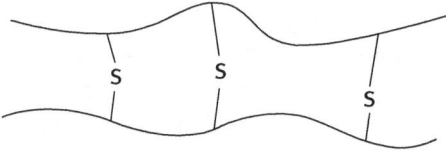

Fig. 1.1: Sulphur bridges in vulcanised rubber.

In 1845, R.W. Thomson invented the pneumatic tyre, the inner tube and even the textured tread. In 1850, rubber toys were being made as well as solid and hollow balls for golf and tennis. The invention of the velocípede by Michaux in 1869 led to the reinvention of the tyre, because Thomson's discovery had been forgotten. In 1888, the use of air-filled rubber tube on his son's bicycle by the Scottish John Dun lop was the start of a new era. In 1895, Michelin adapted the tyre to the automobile. Since that moment rubber has held an outstanding position on the global market, playing a leading role in the modern civilisation. The vulcanisation discovery has been a crucial break through and the beginning of the modern rubber industry in the world, as demonstrated by the large increase in the production of natural rubber from 750 tonnes in 1850 to 6,000 tonnes already in 1860.

As rubber became an important raw material in industry, chemists soon became curious to learn more about its composition in order to synthesise it. In the second half of twenty-first century, some studies between 1879 and 1882 discovered that rubber is an isoprene polymer. Bouchardt revealed how to polymerise isoprene. In 1904, Mote and Mathews in London investigated the reinforcement of rubber by carbon black. At the beginning of the nineteenth century, especially Russia and Germany made many efforts to synthesise rubber, but the resulting products were unable to compete with natural rubber yet. It was only during World War I that Germany, pressured by circumstances, started a systematic investigation to develop synthetic rubber. This was the springboard for the massive development of the synthetic rubber industry all over the world for producing elastomers, as described in Section 1.1.2.

1.1.1 Natural rubber

Natural rubber is obtained through the coagulation of the latex (consisting predominantly of cis-1,4-polyisoprene) produced by certain plants, particularly the Brazilian rubber tree, named *H. brasiliensis*, native to Amazonia [2].

First, a diagonal incision is made in the bark of the tree and the exuded latex from the cut is collected in a small cup. The gathered latex is strained, diluted with water and treated with acid to allow the suspended rubber particles within the latex to coagulate. The obtained material is pressed between rollers to form thin sheets that are air dried and then are ready for the shipment.

Since the second half of the nineteenth century until 1920s, natural rubber underpinned one of the most important development booms in Brazil. At that time, the Industrial Revolution was expanding rapidly as the world lived through a time of prosperity and discoveries (i.e., automobiles, trams, telephones, electric light and other innovations) that were reflected in all sectors. Thanks to its multiple applications, particularly in the expanding automobile industry, rubber produced from the latex tapped from the trees became a product demanded worldwide. Because there was no lack of rubber trees in the Brazilian Amazon, this brought a boom to Northern Brazil, which at that time was one of the poorest and least inhabited parts of that country. Manaus, the capital of the Amazonas, becomes the economic heart of Brazil. Thousands of immigrants flowed in, invading the forest to tap the latex and turn it into rubber.

The Brazilian government believed that tapping the rubber trees would ensure the presence of Brazilians in the Amazonia region, guaranteeing the national sovereignty over this largely unpopulated area. It opted for geopolitics represented by human settlements instead of geoeconomics that could have produced higher gains.

The output of Amazonia reached 42,000 tonnes a year of natural rubber, with Brazil dominating the global market. This euphoria lasted through to 1910, when the situation began to change: rubber exports began to appear on the market from British Colonies, and Brazil was unable to withstand this fierce competition. This fact was related to a particular event. In 1876, British smuggled out rubber tree seeds from Amazonia to the Botanical Gardens in London. Through grafting, more resistant tree varieties were developed. They were later sent to the British Colonies in Asia where massive rubber plantations were established, particularly in Malaysia, Ceylon and Singapore.

The production techniques in Brazil and Asia were completely different, becoming a significant factor in the business development due to these plantations. The rubber trees of Asia were set one very close to the other ones (only 4 m apart). On the contrary, in Amazonia it was sometimes necessary to walk miles between one tree and the next one, limiting the amount of collected latex and thus increasing its price. As a consequence, the well-organised plantations of the Far East resulted in a noteworthy increased productivity, making them more competitive. This relative immobility in the change of the plantation methods costed to Brazil: exports shrank since Brazil could not withstand the competition of Asian rubber, tapped at lower prices. Consequently, production began to drop, bringing the decades of boom to an end for much of Northern Brazil.

In the late 1920s, Brazil was helped to rise again the rubber production by an unexpected partner, the US industrialist Henry Ford, who developed in his vehicle industry a new scheme called 'production line', which was to change the face of industry forever. In order to overlap the leadership of the British Colonies in Southeast Asia on the rubber market, Ford planted no less than 70 million rubber trees in the Para State. Unfortunately, the Ford project succumbed to the hostile environment of the Amazon rainforest and was abandoned, posting huge losses.

From that period, Asia started to dominate the global market of natural rubber, the precious raw material for making tyres, with over 90% of the output. Overtime, significant changes redistributed the production among the main competitors. Accounting for one-third of global output in 1985, Malaysia fell back due to alterations in its production profile towards non-agricultural investments. Therefore, its position as the world's largest natural rubber producer went to Thailand. Based on advantages in terms of available land and labour, Indonesia has maintained a significant share in the global output since the 1980s. Other countries have been successfully deploying their low-cost labour forces and easily available lands to expand in this sector, particularly India and China. Nowadays, more than 90% of all natural rubber comes from rubber plantations of Indonesia, Malay Peninsula and Sri Lanka. The importance of rubbers can be judged from the fact that global revenues are forecast to rise to US $56 billion in 2020.

1.1.2 Synthetic rubber

As already introduced, the decisive role that rubber has played in the development of modern civilisation prompted much interest in discovering its chemical composition in order to synthesise this product artificially [1, 3]. Through several researches, the tyre industry saw the possibility of breaking away from the grip of the natural rubber plantations, as discussed earlier.

The efforts to make synthetic rubber started even before the Staudinger formulation of the macromolecular concept, particularly in the period around World War I in Germany and some years later in the USA.

Sodium polymerised butadiene was produced in Germany (that restarted the synthetic rubber research programme in 1926) as Buna rubber and in the USSR as SK rubber. In the 1930s, Germany developed the emulsion copolymerisation of butadiene with styrene (Buna S), whereas sodium polybutadiene continued as the principal general-purpose synthetic rubber in the Soviet Union. These were the inputs for the development of many synthetic rubbers made in the succeeding decades.

World War II highlighted the importance of rubber as a raw material as witnessed by a crucial historical episode that altered the scenario for this market. On December 7, 1941, the USA entered the War after the attack on Pearl Harbor. Three months later, Japan invaded Malaysia and the Dutch East Indies, giving to the Axis the control over 95% of world rubber supplies, plunging the USA into a crisis.

Each Sherman tank contained 20 tonnes of steel and half a tonne of natural rubber. Each warship contained 20,000 rubber parts. Rubber was used to coat every centimetre of wire used in every factory, home, office and military facilities throughout the USA. There was no synthetic alternative. Looking at all the possible sources, at normal consumption levels, the nation had stocks for around 1 year, including the arms segment. In this framework, the response of Washington was rapid and dramatic: 4 days after Pearl Harbor, the use of rubber in any product that was not essential to the war drive was banned. For example, the speed limit on US highways fell to 35 miles/h, in order to reduce wear and tear on tyres countrywide. The largest recycling campaign ever recorded in history started, ensuring the success of the Allies through 1942.

In the meantime, due to these circumstances, a straight order was sent to all US chemists and engineers to develop a synthetic rubber industry, also suitable for tyres. The nation survival depended on its capacity to manufacture over 800,000 tonnes of products that had barely begun to be developed. If the synthetic rubber program failed, the capacity of the USA to fight the war would be blunted. Very quickly, the research efforts gave birth to a new synthetic rubber: the styrene–butadiene rubber (SBR), then called GR-S. Its production began in a US government plant in 1942. Over the next 3 years, government-financed construction of 15 SBR plants brought the annual production to more than 700,000 tonnes of synthetic rubber.

After the war time and the introduction of SBR, a wide variety of other synthetic rubbers has been developed. In the early 1960s, production of natural rubber was surpassed by that of synthetic elastomers, thanks to massive investments in this research field. Production technology was heavily concentrated in long-established global enterprises such as DuPont, Bayer, Shell, Basf, Goodyear, Firestone, Michelin, EniChem, Dow and Exxon. Since 1990, two-thirds of world rubber production consisted of synthetic varieties.

Both the characteristics and properties of synthetic rubber make them useful in almost all economic sectors, such as automotive, footwear, civil construction, plastics and medical applications, all of crucial importance in the daily life of our society (Tab. 1.1). SBR and BR varieties are the most widely consumed types of synthetic rubber due to their widespread application in tyre industry.

About 90% of today's rubber products contain sulphur as the main vulcanisation system. The spectrum of vulcanisation systems currently used today comprises also peroxides, sulphur donors and special cure chemicals. Virtually, all of the stages in rubber manufacturing, including mastication, compounding, milling, vulcanisation and finishing, are the same today as they were at the beginning of the twentieth century, although the machinery has been continuously refined and improved.

Tab. 1.1: Main types and applications for synthetic rubbers.

Rubber	Asphalt	Foot wear	Adhesives	Technical goods	Tyres	Plastics
Styrene–butadiene (emulsion) – eSBR	–	X	X	X	X	–
Styrene–butadiene (solution) – sSBR	X	X	X	X	X	–
Polybutadiene – BR	–	X	–	X	X	X
Nitrile rubber – NBR	–	X	–	X	–	X
Ethylene–propylene diene monomer – EPDM	X	–	–	X	X	X
Butyl rubber – IIR	–	–	X	X	X	–
Polychloroprene – CR	X	X	X	X	–	–
Thermoplastic elastomers – TPE	X	X	X	–	–	X
Latex	X	X	–	X	–	–

1.2 Elastomers

The term *elastomer*, derived from 'elastic polymer', is often used interchangeably with the term *rubber*, although the latter is preferred when referring to compounds and, consequently, to vulcanisates [1, 3, 4].

An elastomer is a polymer that exhibits high elasticity and reversible elongation up to 1,500%: it can be stretched several times without breaking, and upon release of the stress, it immediately returns to its original length. So, elastomers can exhibit rapid and large reversible strain in response to a stress and show almost no creep. For example, an ordinary elastic band can be stretched up to 15 times its original length and then can be restored to its original size. Elastomers are incompressible in bulk and have a Poisson's ratio close to 0.5.

Elasticity is derived from the ability of the long macromolecular chains to reconfigure themselves to distribute an applied stress. When the elastomeric chains are stressed in a preferred direction, chains elongate (Fig. 1.2). When the stress is removed, they will return to the initial configuration (elastic range), thanks to the presence of covalent cross-links, which prevents the slipping of the macromolecular chains. Without any cross-link, the applied stress would result in a permanent deformation.

Therefore, an elastomer can be defined by its mechanical response when a force is applied. It undergoes an immediate, linear and reversible (elastic) response, showing a mechanical analogy with a spring according to the Hooke's law. Conversely, the time-dependent irreversible (viscous) response is in accordance to the dashpot model. An ideal elastomer exhibits only the elastic response. In real elastomers, beside the elastic deformation (change of shape or size that lasts only as long

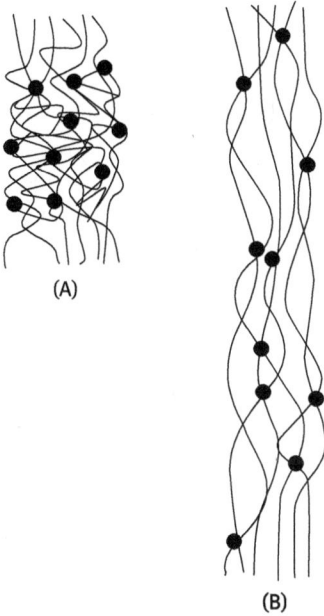

Fig. 1.2: Schematic representation of an unstressed elastomer (A) and the same elastomer under stress (B), in which dots represent the cross-links.

as the deforming force is applied and after it disappears), also a viscous response at higher strains is exhibited. Therefore, the non-linear, time-dependent mechanical response of elastomers is identified as viscoelasticity according to the coupling of the spring and the dashpot parallel model. In few words, an elastomer is a viscoelastic polymer (showing both viscosity and elasticity) characterised by very weak intermolecular forces, low Young's modulus and high failure strain compared with other materials, as shown in Fig. 1.3.

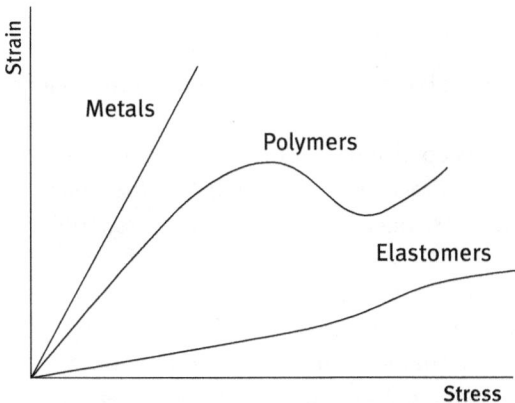

Fig. 1.3: Typical stress–strain curves for metals, polymers and elastomers.

Cross-links are essentially formed by the addition of sulphur, which at high temperatures forms the sulphidic bridges between different macromolecular chains (Fig. 1.1) in the presence of compounds known as accelerators. Once cross-linked, the elastomer cannot be reprocessed or recycled. The most used unsaturated rubbers that can be cured by sulphur vulcanisation are synthetic polyisoprene, polybutadiene, polychloroprene (also called neoprene), butyl rubber, SBR, nitrile rubber and hydrogenated nitrile rubber. Some unsaturated rubbers can also be cured by non-sulphur vulcanisation, for example, by peroxides. Conversely, saturated rubbers cannot be cured by sulphur vulcanisation (i.e., EPDM, epichlorohydrin rubber, polyacrylic rubber, silicone, fluorosilicone, fluoroelastomers, perfluoroelastomers and chlorosulphonated polyethylene).

It is possible for a polymer to exhibit elasticity that is not due to covalent cross-links, but instead to thermodynamic reasons as in the case of thermoplastic elastomers. Physical reversible cross-links can be due to a phase-separated block copolymer structure where a disperse phase in a flexible matrix is glassy (high T_g) or crystalline (Fig. 1.4). Thermo plastic elastomers can be moulded, extruded and reused like thermoplastic materials, while still having the typical elastic properties of elastomers, as explained in Chapter 5.

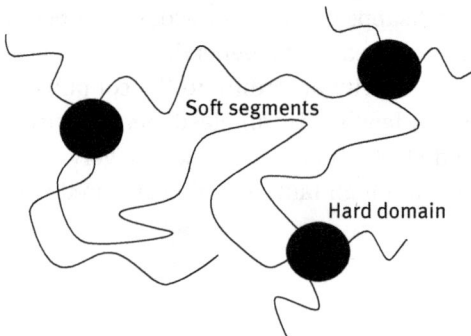

Fig. 1.4: Schematic structure of thermoplastic elastomers.

For having a rubbery behaviour, some main conditions must be verified in the macromolecular structure of any elastomer [5]:

- *Long, highly flexible chains.* Elastomers consist of flexible polymeric chains able to alter their arrangements, conformations and extensions in the space in response to an imposed stress. Only the macromolecular chains are so highly flexible to allow any conformational change necessary for having both reversibility and immediate elastic response. An unstrained elastomer exists in a random coil structure. A random coil is more thermodynamically stable compared with the fully extended chain because there are an infinite number of random coils, while there is only one fully extended chain. As strain is increased, the macromolecular chains uncoil to the limiting linear structure. Steric hindrance to uncoiling should be minimal so that elastomers are unlikely to have bulky pendant groups or rigid intra-chain groups. This is why most common elastomers consist of simple hydrocarbon high macromolecular chains and stiff groups such as benzene, bulky side-chains such as isopropyl, polar groups and hydrogen bonding groups are not desirable if a polymer has to be an elastomer. In addition, it is necessary that all the configurations are accessible. Upon deformation, the number of configurations available to a chain decreases and the resulting decrease in entropy gives rise to the retractive force. This means that the rotation (the result of thermal agitation) must be relatively free about a significant number of bonds joining neighbouring skeletal atoms in the polymer backbone.

– *Network structure.* The macromolecular chains must be joined by cross-links, as illustrated in Fig. 1.2. The resulting network structure is essential to avoid the chains to permanently slip by one another, which would result in flow and thus in irreversibility. Cross-links may be chemical bonds or physical links, such as the glassy domains in the multi-phase block copolymers (Fig. 1.4).

– *Weak interchain interactions.* The macromolecular chains must be free to move reversibly past one another. Thus, the intermolecular attractions known as secondary or van der Waals forces, which exist between all macromolecules, must be small.

– *Glass transition temperature.* Elastomers must be well above their glass transition temperature so that a considerable segmental motion is possible.

– *Amorphous structure.* Elastomers must have an amorphous arrangement and no crystallinity at least at low strain since any regular crystal structures are unable to contribute to the elastomeric properties. Irregularity in the molecular structure is essential to prevent crystallinity. For example, a fully saturated hydrocarbon polymer can be elastomeric if all the substituents are in the atactic configuration, or if it is a random copolymer where the segments cannot cocrystallise.

In few words, elastomers are highly flexible, elastic, amorphous and random-oriented polymers in which the polymeric chains are joined by chemical bonds that allow acquiring a final cross-linked structure.

1.3 Rubber applications

The peculiar properties of rubber materials have led them to be used in a vast range of applications, which can be divided into two broad categories: sealing and non-sealing [1, 6].

Seals are produced using virtually all types of elastomers with a wide range of reinforcements, fillers, additives and cross-linking technology. The inherent elastic properties of rubber make it a natural choice for sealing applications. Indeed, the versatility of elastomer systems offers value in the 'fine-tuning' of requirements to specific service conditions. However, some properties of elastomers, such as high compression set resistance but poor dynamic performance or high heat resistance but low chemical resistance to specific fluids, could hamper the use for well-defined sealing applications.

Moulded or machine shaped seals are designed in geometry and formulation to resist pressure, motion and environment to which they are exposed in service. In some cases, where the elastomer material is not inherently strong enough to withstand the environment in which it is exposed, additional components can be bonded or included in the seal design to increase the elastomer performance (i.e., engineering plastics or metallic components).

Rubber materials are also widely employed in non-sealing applications such as elastomeric belts used in drive systems and power transmission. Flexible hoses used to transfer fluids from one point to another one or to transmit energy, such as in hydraulic applications, are also typically made by reinforced elastomers and may be multi-layered by design. Elastomers are also employed in personal protection and diving products, including face masks, nasal units, fixing straps, neck seals, ankle and wrist seals, regulator valves and mouthpieces.

Rubber materials also have many applications in civil engineering, for example, for mounting structures to reduce the effect of external noise, vibration or seismic forces, for accommodating thermal movements (bridge bearings, expansion joints, pipe couplings, etc.), for acting as a barrier to water (water stops, plant linings, tunnel gaskets) and for a wide variety of other uses including roofing membranes, rubberised asphalt, rail pads, inflatable formers and concrete texturing.

The automotive sector is the largest user of elastomers, with tyres being by far the widest application by volume. Another key use is in the suspension systems, where the components can be designed with very specific dynamic properties.

1.4 Thermodynamics of elasticity

Elasticity is the reversible stress–strain behaviour exhibited by rubber-like materials in a unique and extremely important manner [7, 8]. The reversible large deformability of an elastomer is reminiscent of a gas. In fact, the term *elastic* was first used by Robert Boyle (1660) in describing a gas: 'There is a spring or elastical power in the air in which we live.' In 1805, long before anything was known about the molecular structure of rubber occurring later, its mechanical properties were investigated by Gough, who reported that the length of a rubber sample held under constant stress decreased as its temperature was increased. He demonstrated that heat was evolved as a result of the adiabatic extension. This is in sharp contrast to the familiar response of expansion shown by other solids and liquids.

The thermodynamic study of rubber elasticity was started on the theoretical side by Lord Kelvin in 1857 and on the experimental side by the very precise work of James Prescott Joule in 1859, through measurements of force and specimen length at different temperatures [9]. Both discovered the thermoelastic effect, demonstrating that rubber exhibits predominantly an *entropy-driven* elasticity. Lord Kelvin experimentally observed that a stretched rubber sample, subjected to a constant uniaxial load, contracts reversibly and gives out the heat reversibly. Thus, he concluded that the tension in a rubber sample arose from the heat motion of the constituent particles in rubber. In this way he defined the thermodynamic laws of elasticity.

At that time, any attempt to specify better the nature of the constituent particles of rubber was unsatisfactory; thus, it was not until after 1930, by which time the macromolecular character of the rubber molecules was completely understood and accepted that the development of a quantitative kinetic theory of the rubber-like elasticity became possible. Indeed, the instantaneous deformation occurring in rubber during stretching could be explained in terms of the high segmental mobility and the rapid changes in the conformation of macromolecular chains. The energy barriers between the different conformational states must be small compared to the thermal energy. Due to this behaviour a number of spectacular phenomena can occur: stiffness increases with increasing temperature and heat is reversibly generated on deformation.

Thermodynamics of gases predicts that when a gas is heated under constant volume its pressure increases. In the thermodynamics of elastomers there is an analogous prediction: when a stretched elastomer is heated its length will decrease. This can be easily observed with an ordinary rubber band. Stretching a rubber band causes to release heat, while discharging it, after it has been stretched, will lead to absorb heat, causing its surroundings to become cooler.

Real elastomers display hysteresis upon stretching or relaxation. The hysteresis in the stress–strain curves makes that the elongation and contraction curves do not coincide. The area between them represents the energy lost during each cycle, which manifests as heat.

The application of thermodynamics to the elastomeric deformation should require that the macromolecules rapidly equilibrate and comply with Hooke's law over the entire range of strain being considered [10]. Such kind of elastomer is defined as *ideal elastomer*, by analogy with an ideal gas. However, like most gasses, most elastomers are not ideal: real elastomers do not behave completely reversibly. Indeed, at high deformations the elastomeric chains become fully extended between the chemical or physical cross-links and the stress–strain response becomes nonlinear (viscoelasticity). The molecular entanglements prevent the free molecular uncoiling and, therefore, cause deviations from a linear elastic response. In few words, deviations from ideality are caused by the finite size of the macromolecular chains and the limited free volume available for molecules to occupy.

1.4.1 Theoretical approach

As stated earlier, elastomers extend and contract by conformational change from a compact random coil to extended chains [7, 8]. The entropy of an elastomer is the disorder that is increased as the elastomer is contracted since the number of possible conformations of each macromolecule approaches infinity. Since the random coil can have many possible conformations, this results in high entropy, whereas a fully extended chain can only have one conformation resulting in low entropy. Therefore, stress acting on the rubber network will stretch out and orient the chains between the cross-links, decreasing the entropy of the system to a minimum. The entropy decrease on stretching is due to the fact that the elasticity of rubber is predominantly entropy driven. Also the retractive force that allows the chains to adopt higher entropy conformations releasing the applied stress has an entropic nature in rubbery materials.

When a compressed gas is allowed to expand, its temperature will decrease. A stretched elastomer shows the same behaviour [11]. Therefore, an elastomer behaves somewhat like an ideal monatomic gas, in as much as (to good approximation) elastic polymers do not store any potential energy in the stretched chemical bonds or any elastic work done in stretching molecules, when work is applied upon them. All work done on the rubber is thus released (not stored) and appears immediately in the polymer as thermal energy. Since the ability of an elastomer to do work depends (as with an ideal gas) only on entropy-change considerations, and not on any stored energy (i.e., potential) within the polymer bonds, it is necessary to extend the thermodynamic treatment of the elasticity, sharing the elastic force, originated from changes in both the conformational entropy and in the internal energy, into entropic and energetic contributions. The thermodynamic approach reported below tries to explain the elastic behaviour of elastomers differentiating between the entropic and energetic contributions to the elastic force.

Rearranging $\Delta G = \Delta H - T\Delta S$, where G is the free energy, H is the enthalpy and S is the entropy, $T\Delta S = \Delta H - \Delta G$ is obtained. Since stretching is non-spontaneous, as it requires external work, $T\Delta S$ must be negative. T is always positive (it can never reach absolute zero), thus ΔS must be negative, implying that the rubber in its natural state is more entangled than when it is under tension. Thus, when the tension is removed, the reaction becomes spontaneous, leading ΔG to be negative. Cooling effect must result in a positive ΔH, so ΔS will be positive there.

The change in internal energy dU during stretching an elastic body is given by the first law of thermodynamics:

$$dU = dQ - dW \tag{1.1}$$

where dQ is the element of heat absorbed by the system and dW is the element of work done by the system on the surroundings.

For the second law of thermodynamics, in a reversible process entropy is a function of state and it is given by

$$dQ = TdS \tag{1.2}$$

The entropy in an adiabatic system can never decrease, but it increases during an irreversible process remaining constant during a reversible process.

The work dW can be expressed as the sum:

$$-dW = -pdV + fdL \tag{1.3}$$

where p is the equilibrium external pressure, dV the volume dilation accompanying the elongation of the elastomer, f the equilibrium tension and dL the length change. Thus, the change in internal energy dU is given by

$$dU = TdS - pdV + fdL \tag{1.4}$$

where dS is the differential change in entropy, pdV is the pressure volume work and fdL is the work done by deformation.

At constant pressure, the enthalpy change is

$$dH = dU + pdV = TdS + fdL \tag{1.5}$$

A deformation dL at constant pressure and temperature induces a retractive force f expressed as:

$$f = \left(\frac{\partial L}{\partial H}\right)_{T,p} - T\left(\frac{\partial L}{\partial S}\right)_{T,p} \tag{1.6}$$

This is one of the forms of the thermodynamic elastic equation of state.

A basic postulate of the elasticity of amorphous polymer networks is that the stress exhibited by a strained polymer network is assumed to be entirely intramolecular in origin and, consequently, the intermolecular interactions play no role in deformations at constant volume. The energetic and entropic contributions to the force in the intramolecular process of stretching chains can be evaluated in experiments at constant volume where there is no other energetic contribution resulting from changes in (intermolecular) van der Waals forces (no enthalpic contribution). As a consequence, an equation similar to eq. (1.6) is obtained for the elastic force measured at constant volume:

$$f = \left(\frac{\partial L}{\partial U}\right)_{T,V} - T\left(\frac{\partial L}{\partial S}\right)_{T,V} \tag{1.7}$$

1.4.2 Statistical mechanical theories of rubber elasticity

Statistical mechanical models based on thermodynamic, statistical mechanics and continuum mechanical considerations have been useful in describing the stress–strain behaviour of rubbers [8]. Theories explaining rubber elasticity have passed through three distinct phases: the early development of a molecular model based on experimental observations; then, the generalisation of this approach by means of symmetry considerations based on the assumption that continuum mechanics is independent of the molecular structure; and most recently, a critical reassessment of the basic statements on which these former theories are founded.

The first molecular-based statistical mechanical theory, called as the *affine network model*, was developed by Wall, Flory and Rehner, with the simple assumption that the chain segments of the network deform independently and on a microscopic scale in the same way as the whole sample (affine deformation). The macromolecular chains are firmly embedded in a deformable continuum [12]. Conversely, James and Guth allowed in their *phantom network model* a certain fluctuation (free motion) of the cross-links around their average affine deformation positions. This network comprises volumeless chains able to pass through one another [13].

These two theories describe two extremes in the relationship between stress and strain, with the *affine network model* giving an upper bound modulus and the *phantom network model* the lower bound. It seems thus that the fluctuations of the junctions are suppressed at low extensions and that further extension makes the junctions more mobile to fluctuate. Both models assume that the chains of the network behave like phantom chains, meaning that the dimensions of these chains are unperturbed by excluded volume effects.

The *affine network model* assumes that cross-links have a specified fixed position in the space exactly defined by the specimen's deformation ratio. However, the chains between the junction points are free to take any of the great many possible conformations. In the framework of the *affine network model* an equation relating stress and strain in rubber has been defined, starting from the following assumptions: (i) the chain segments between cross-links can be represented by the Gaussian statistics of phantom chains; (ii) the free energy of the network is the sum of the free energies of the individual chains; (iii) the positions of the cross-links are changed precisely according to the macroscopic deformation; (iv) the unstressed network is isotropic; (v) the volume remains constant during deformation; and (vi) all different conformation states have the same energy.

The Gaussian distribution function for the end-to-end distance expresses the probability of finding the end of a chain at a certain position (x,y,z) with respect to the other chain end found at $(0,0,0)$. In other words, the number of conformations a chain can take when the chain ends are in $(0,0,0)$ and (x,y,z) can be statistically evaluated. Starting from this assumption, it is then possible to calculate the Helmholtz free energy ΔA (the free energy associated with constant volume) for a single chain, adding each contribution from all individual chains of the network. The stress–strain equation may be obtained as the derivative of the Helmholtz free energy with respect to length:

$$\Delta A = \left[NkT\left(\lambda_x^2 + \lambda_y^2 + \lambda_z^2 - 3 \right) \right]/2 \qquad (1.8)$$

where N is the number of chains of the network and λ_x, λ_y, λ_z indicate the dimensional variations on each dimension (being λ the stretch ratio between the deformed and undeformed length). This equation is general and not restricted to any particular state of strain.

When the stress applied to a cross-linked coil is released, deformation is a constant volume process. In that case, the Helmholtz free energy should be minimised, assuming that the internal energy is independent by deformation. As a result, the entropy is maximised.

The *phantom network model* permits fluctuations of the junction points around the average positions prescribed by the macroscopic deformation, so the chains between the cross-links can take any of the many possible conformations. The derivation of the free energy–strain equation is here more complicated since the chains of the network are coupled and the probability function for the network is the product of the probability functions of the individual chains. As the average positions of the junctions are deformed affinely, the average fluctuation of any given junction is independent on strain and can be described by a Gaussian probability function. The forces exerted by the network are the same whether any given junction is treated as free, or as fixed at its most probable position. These findings justify much of the assumptions made in the *affine network model* and there is no surprise that the free energy equations and the derived stress–strain equations are similar for the two models. Indeed, the free energy–extension ratio expression derived by James and Guth is

$$\Delta A = \left[(1 - 2/\psi)NkT\left(\lambda_x^2 + \lambda_y^2 + \lambda_z^2 - 3 \right) \right]/2 \qquad (1.9)$$

which for an infinite tetrahedral network ($\psi = 4$) becomes

$$\Delta A = \left[NkT\left(\lambda_x^2 + \lambda_y^2 + \lambda_z^2 - 3 \right) \right]/4 \qquad (1.10)$$

that is precisely half of the value predicted by the *affine network model*.

Both theories fail to predict the stress–strain behaviour at large strain (i.e., $\lambda > 4$ for natural rubber), as demonstrated by the deviation of experimental results from the predicted trend. This led to the development of the Mooney–Rivlin equation in which the elastic constants can be determined by a linear regression of the reduced stress [14–16].

A further important development in the theory of rubber elasticity was the Flory's recognition of the role of intermolecular interactions in modifying the stress in a strain-dependent manner. This idea was incorporated into the *constrained junction*

(CJ) model, where Flory and coworkers introduced a constraint parameter K, which at high extension (no constraints) takes the value 0 (*phantom network*) and at low extensions (infinitely strong constraints and suppression of junction fluctuations) takes the value ∞ (*affine network*) [17, 18]. The value of K increases with the degree of constraints. This development of the fundamental statistical mechanical theories, including an additional adjustable parameter, is thus capable of describing the stress–strain behaviour in the intermediate extension ratio range ($1.2 < \lambda < 2$).

Later, Flory and Erman extended the CJ model, introducing an additional parameter, ξ (>0), to keep into account any difference from the affine distortion, due, for example, to some inhomogeneities in the network structure [19]. A large value of ξ implies that the constraints are alleviated by the extension of the network more rapidly than if the domains would deform affinely. The introduction of this parameter allowed to have the quantitative agreement with the experimental data.

References

[1]   Blow, CM, Hepburn, C. Rubber technology and manufacture. Buttenworths, London, UK, 1982.
[2]   Roberts, AD. Natural rubber science and technology. Oxford University Press, Oxford, UK, 1988.
[3]   Franta, I. Elastomers and rubber compounding materials. Elsevier, Amsterdam, The Netherlands, 1989.
[4]   Sperling, LH. Introduction to polymer science. Wiley-Interscience, Hoboken, New Jersey, USA, 2006.
[5]   Mark, JE, Erman, B, Eirich, FR. The Science and technology of rubber. Elsevier Academic Press, Amsterdam, The Netherlands, 2005.
[6]   Dick, JS. Rubber technology – Compounding and testing for performance. Hanser Publishers, Munich, Germany, 2001.
[7]   Mark, JE. Physical properties of polymers. American Chemical Society, Washington DC, USA, 1984.
[8]   Brydson, JA. Rubbery materials and their compounds. Elsevier Applied Science, London, UK, 1988.
[9]   Meares, P. Polymers. Structure and bulk properties. Van Nostrand, London, UK, 1965.
[10] Simpson, RB. Rubber basics. Rapra, Shrewsbury, UK, 2002.
[11] Elias, HG. Macromolecules – physical structure and properties. Wiley VCH, Weinheim, Germany, 2008.
[12] Kuhn, WJ. Dependence of the average transversal on the longitudinal dimensions of statistical coils formed by chain molecules. Polym. Sci. 1946, 1, 380.
[13] James, HM, Guth EJ. Theory of the increase in rigidity of rubber during cure. J. Chem. Phys. 1947, 32, 669.
[14] Treloar, LRG. The physics of rubber elasticity. Clarendon Press, Oxford, UK, 1975.
[15] Rivlin, RS. The elasticity of rubber. Rubb. Chem. Technol. 1992, 65, G51.
[16] Flory, PJ. Molecular theory of rubber elasticity. Polymer. 1979, 20, 1317.
[17] Flory, PJ. J. Chem. Phys. 1977, 66, 5720.
[18] Flory, PJ. Theory of elasticity of polymer networks. The effect of local constraints on junctions. Statistical mechanics of chain molecules. Wiley-Interscience, New York, USA, 1969.
[19] Flory, PJ, Erman, B. Theory of elasticity of polymer networks. 3. Macromolecules. 1982, 15, 800.

2 Compounding

2.1 Introduction

Basic properties of rubbers are highly dependent not only on the raw polymers used in their manufacture, but also they can be modified by an appropriate addition of well-defined ingredients leading to the formation of the rubbery mixture, that is, the compound [1]. Therefore, the compound preparation, called also *compounding*, is one of the key processes in the rubber industry because both the material characteristics and behaviours largely depend on the quality of mixtures.

Any rubber compound can be designed for specific purposes and applications, tailoring both the nature and the relative amounts of their constituents. Some of them can help to accelerate the cross-linking process, some improve the processability and others permit to attain the highest levels of performance in the vulcanisates, with cost being a secondary issue. In specific applications, compounds may be designed to minimise the cost, adding extenders and/or diluents to reduce the proportion of high-priced components in the mixture. Of course, this inevitably may compromise both mechanical and other properties, but for certain general applications this drawback can be acceptable.

Currently, all new rubber compounds are modifications of some of the existing formulations. However, when the development of a completely new compound is attempted, first the customer requirements must be kept into account to achieve the required balance in the vulcanisate properties. Durability, cost and easy processing must be considered as well.

2.2 Basics on rubber compounding

Three major factors influence the quality of any moulded rubber item: the mould, the process and the rubber formulation [2, 3]. These three factors are interactive, even if the most influential is the latter. In compounding, literally hundreds of variables related both to materials and equipment have to be examined and faced. There is no infallible mathematical formulation to help the compounder. This is why compounding is a difficult task and it may be considered as partially art and partially science.

In plastics, designers can use the material technical sheets coming from the suppliers to determine what material will work best for a given application. Unfortunately, this kind of information is not readily available in the case of rubber. In addition, the plastic moulder purchases materials in a ready-to-process form, and the resulting physical and environmental resistance are controlled by that supplier.

https://doi.org/10.1515/9783110640328-002

In contrast, the rubber compounder purchases ingredients from many suppliers and mix them to form the rubber, having a direct control on both the material and the resulting vulcanisate properties.

Often designers have little experience with rubber and have no idea how to select the best one for their application. For example, in the general literature it is reported that nitrile rubber is rated excellent in resisting to petroleum hydrocarbons. Thus, people who are not skilled could believe that specifying only the hardness value the vulcanisate properties could be controlled. What designers do not realise is that rubber compounders have around 200 different commercial grades of nitrile to choose from, keeping the hardness value constant. They vary by acrylonitrile content, viscosity, chemistry and supplier (equivalent grades usually are not) among other parameters. All of them influence the resulting end-performances of nitrile rubber.

The availability of thousands of elastomers and ingredients provides a compounder an extremely large number of choices, thus increasing the difficulty in selecting the ingredients. Since any ingredient is available under as many as 20 different trade names, a compounder should think not in terms of trade names but in terms of chemical composition and potential reactions. Unfortunately, with some ingredients this is not possible because manufacturers do not disclose their composition.

Since, more than any other material, the environment and the service conditions largely affect elastomers, the selection of the best elastomer for a given application is critical. The extremely wide range of elastomers available to the rubber technologist is both bad and good news: bad because the behaviour and properties of the myriad available elastomers overlap one another and complicate the choice of an individual elastomer or a blend of elastomers that will meet the requirements for the final use, and good because the wide range of available elastomers provides the technologists with materials to meet the requirements for most applications, some of which are extremely demanding.

Besides the quality of the raw ingredients, the kind of the mixer used for the compounding and the quality control in mixing affect both the quality and the performances of the resulting compound. Usually, compounding is done in either a two-roll rubber mill or a Banbury internal mixing mill, as described in Section 2.5.

Tab. 2.1 lists the formulation of a typical rubber compound, consisting of 10 or more ingredients. All compounders use the unit of measure as *parts per hundred* (phr) in any formula. This is a unit of weight that keeps into account the relationship between the elastomer and the other components. In any formulation 100 phr of elastomer are present, allowing a simpler variation into the relative amounts of the other ingredients and an immediate comparison among different compouunds, Indeed, since the cure system reacts only with the elastomer, even when all the

Tab. 2.1: Standard formulation for a generic rubber compound.

Ingredient	Phr
Elastomer	100
Fillers (reinforcing and non-reinforcing)	0–200
Plasticisers	0–40
Activators	0–40
Stabilisers (i.e., antioxidants, anti-ozonants, protective wax)	0–10
Vulcanising agents	0.3–50
Accelerators	0.3–4
Other additives	0–10

other components are changed, the relationship between them will almost remain constant, with a few exceptions.

2.3 Main ingredients in a rubber compound

As mentioned, many selected ingredients are combined to produce a rubber formulation [1, 2], which are briefly described in the following. More details on some of them are reported in Chapters 3 and 6.

2.3.1 Elastomer

The elastomer, or blend of elastomers, is an essential component in determining the main properties of the rubber compound. It is selected to optimise the service performances and the processing requirements of the compound. A number of specialty elastomers available to compounders overcome some of the deficiencies of general-purpose rubbers, such as poor resistance to oil and fuel, high-temperature ageing and poor flexibility at extremely low temperatures.

2.3.2 Fillers

Fillers are primarily added to provide reinforcement and secondly to reduce the cost. They can influence the processing properties as well. They fall into three basic categories: reinforcing, semi-reinforcing and non-reinforcing. The most popular reinforcing fillers are carbon blacks, categorised primarily by the particle size: they become more reinforcing with the decrease in the particle size. Non-reinforcing fillers (i.e., soft clay, calcium carbonate and talc) have a larger particle size and do not interact with the polymeric matrix as the reinforcing fillers. Thus, they are mainly added to reduce cost.

2.3.3 Curatives (vulcanising agents)

Sulphur, peroxides, resins and metal oxides are typically used as vulcanising agents to form the cross-linked network. Their use varies based on the type of elastomer. The use of sulphur alone leads to a slow vulcanisation process, and so accelerators are added to increase the cure rate. Peroxide cures give good thermal stability due to the short length of the cross-links between the polymer chains.

2.3.4 Accelerators

In a formulation it is common to use more than one accelerator to speed up the cure. Peroxide-cured materials often use a co-agent along with the peroxide that acts as an accelerator and can modify the physical properties of vulcanisates.

2.3.5 Activators

In most sulphur cured rubbers, zinc oxide and stearic acid are added to help to initiate the curing. In other rubbers, different chemicals that assist the cure in an indirect way are added as well.

2.3.6 Retarders or inhibitors

During mixing and further processing in a calender, extruder or moulding machine, rubber is continuously subjected to heat, which can result in premature curing (scorching) or pre-curing. To prevent this, retarders or inhibitors are mixed with the compound to impart a larger scorch safety.

2.3.7 Plasticisers

Plasticisers, called also softeners or lubricants, need to be compatible with the elastomer, since they decrease its viscosity. Besides fillers, plasticisers (i.e., mineral oils and paraffins) play the biggest quantitative role in building a rubber compound. They improve the flowability of rubber during processing and the filler dispersion. Again, they influence the physical properties of vulcanisates at low temperatures and may also work as processing aids. Plasticisers can cause problems when a vulcanisate is subjected to thermal cycling and/or to certain solvents, as they can leach out at high temperatures.

2.3.8 Pre-dispersed ingredients

Some types of ingredient that are difficult to disperse (i.e., certain accelerators and antioxidants) can be used as pre-mixed in an inert polymer in concentrations about 75–80%.

2.3.9 Processing aids

Processing aids are added in a small amount to improve the performance of the compound in processing, thus helping the filler dispersion. Examples include physical peptisers, lubricants, silicone-modified processing additives and anti-stick agents.

2.3.10 Flame retardants

To improve the flame resistance, a number of chemicals may be added to the compound, either inorganic or organic, such as antimony trioxide, zinc borate, aluminium hydroxide and chlorinated paraffins.

2.3.11 Odorants and deodorants

This class of compounding ingredients was more common in the days when natural rubber was the main used rubber worldwide to overcome its distinctive smell. Many of the synthetic rubbers have their own distinct aroma, and often this has to be masked to make the final items acceptable to the users, using specific organic deodorants mixed in the compound formulation.

2.3.12 Peptisers

Peptisers act as chain-terminating agents during the rubber mastication. They significantly reduce the time required to lower the rubber viscosity to a workable level, thereby reducing both the power consumption and the mixing time. Examples of peptisers are pentachlorothiophenol, phenyl hydrazine, certain diphenyl sulphides and xylyl mercaptan. They are most effective in natural rubber, styrene–butadiene rubber and polyisoprene, but are relatively ineffective in other synthetic rubbers. Each peptiser has an its optimum loading in every compound for reaching the maximum efficiency.

2.3.13 Pigments

Although most rubber compounds are black, due to the widespread use of carbon black as filler, inorganic or organic pigments are frequently added. Inorganic pigments are often too opaque to provide the desired colour; they are insoluble and thus cannot bloom. Organic pigments generally give brighter shades but are more sensitive to heat and chemicals and are also relatively expensive. They can also fade badly in long-term exposure to sunlight.

2.3.14 Anti-degradants

These materials are added to inhibit the attack by oxidation and ozone, thus improving the long-term stability of rubber. Ozone, even at concentrations of only several parts per hundred million, attacks general-purpose elastomers that easily crack leading to the product failure. The incorporation of an effective anti-ozonant in those compounds inhibits or prevents the ozone cracking. However, some anti-ozonants, such as waxes, can bloom onto the surface, giving similar drawbacks of plasticisers and processing aids.

2.3.15 Tackifiers

Natural rubber displays a phenomenon known as 'natural tack', that is, the ability of two materials to resist to the separation after being in contact for a short time under a light pressure. This is a consequence of interpenetration of molecular chain ends followed by crystallisation. Tackifiers are low-molecular weight compounds (i.e., rosin derivatives, polyterpenes, pine tar, coumarone resins, petroleum resins and non-reactive phenolic resins) introduced in the compound to enhance the surface tack of uncured elastomers. They are less compatible than a plasticiser, but more compatible than a filler.

2.3.16 Blowing agents

Blowing agents are used in the manufacture of sponge rubber, which show a cellular form in which the macromolecular matrix contains gas filled cells that may or may not be intercommunicating. The most widely used chemical blowing agent are azodicarbonamide, hydrazide derivatives and sodium bicarbonate, which are incorporated in rubber, since they decompose at some stages of the processing operations to yield volatile reaction products such as N_2, CO_2, H_2O, NH_3 and so on, thus forming the cellular structure. Physical blowing agents liberate gases as a result of physical processes such as evaporation or desorption at elevated temperatures or reduced pressures. This class includes mostly volatile liquids, for example, freons, aliphatic hydrocarbons and powdered solid carbon dioxide.

2.4 The compounding process

The compounding goal is to produce a rubber in which all the ingredients are well dispersed and sufficiently distributed so that it will cure efficiently giving the required properties for the end applications [1–3]. Before starting the compounding process, the formulation, what equipment to use, time, speeds, pressures and temperatures must be set up to minimise labour, energy and equipment cost. Because of the viscoelastic nature of elastomers, high-energy machinery like mixing mills and internal mixers (also called masticator) are necessary for allowing the compounding process.

Two different ways of mixing can be followed during the first step of the compound preparation, before the addition of vulcanising agents, recognised as *conventional mixing* and *upside-down method*, respectively. In the *conventional mixing*, the filler incorporation into the polymeric matrix occurs at first to avoid the premature vulcanisation. Fillers are added in one or more stages depending on their nature and amounts. The *upside-down method* uses the reverse practice. Primarily the fillers are fed, then softeners are mixed and lastly the rubber is added into the machine. This method is not very useful for mixtures containing high-activated fillers (i.e., high structural carbon black and active silica) or high content of soft mineral fillers and oils together with high viscosity polymer. On the other hand, some polymers such as EPDM, which do not need the initial softening, can be mixed following the *upside-down method*.

According to the *conventional mixing* method, first the polymer is fed into the internal mixer or an open mill. As a consequence of the frictional forces occurring between the machine components and the viscoelastic material, some heat is generated. The raw polymer undergoes a structural change passing from the elastic to the plastic state. As a consequence, this first mixing step in compounding is defined as plasticising. The large deformations and the subsequent relaxations occurring in the rubber domains allow the next incorporation of fillers aggregates between the rubber layers. Longer storage time of the polymer in the masticator during this phase results in an increase in the viscosity and, consequently, leads to unstabilisation. If plasticising is performed incorrectly, a decrease of the fatigue and ageing resistance can be observed in the vulcanisates, since the macromolecular chains are broken, the molecular weight is reduced and, consequently, the mechanical behaviour is generally worsen.

The relationship between the plasticising efficiency and the temperature is shown in Fig. 2.1. A decrease in the efficiency of the mechanical tearing of macromolecules is observed on the left-hand side, whereas a rising effect of heat, leading

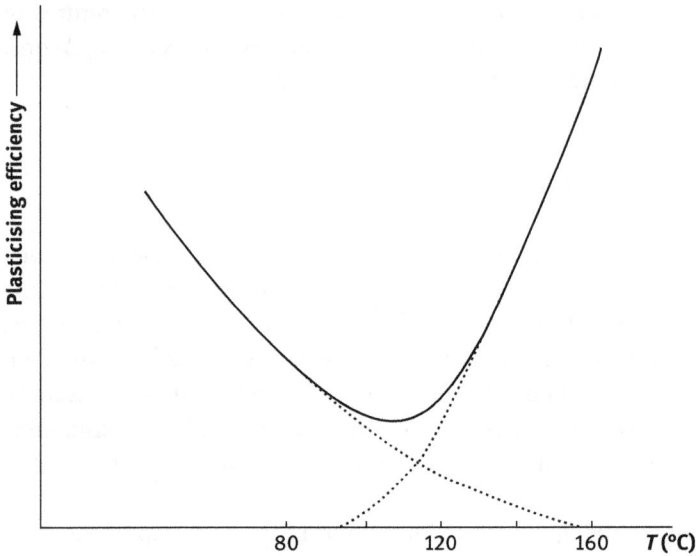

Fig. 2.1: Influence of temperature on the plasticising effect during the first step of rubber compounding in the *conventional mixing*.

to the polymer degradation, is shown on the right-hand side. The maximum plasticising efficiency is achieved under 55 °C or over 140 °C, with the best plasticising efficiency observed at 115–120 °C. Lower temperatures are used for calendering; higher temperatures are typical for internal mixers.

Some factors influence the quality and the efficiency of the plasticising process in the internal mixer, such as the masticator size, the geometry of rotors, the dimensional ratio between the rotor and the chamber wall, the machine filling, the rotor speed, the working temperature and time. For example, it is important to optimise the chamber volume so that the shear on the compound is maximised.

After plasticising and the subsequent stabilisation, the next step, called blending, involves the addition and dispersion of fillers, including carbon black. The degree to which the filler has to be dispersed depends on the quality requirements of the compound: the higher the degree of dispersion, the better the properties. Generally speaking, two different routes may be followed to reach both the optimal distribution and dispersion, as shown in Fig. 2.2. Route 1 allows to reach a system with well-distributed but poorly dispersed additive; it entails lower viscosity than route 2. Thermodynamics studies suggest that route 1 requires more energy than route 2, so the second one is energetically favoured.

Generally, the most part of fillers are automatically weighed and fed directly into the mixer for having high accuracy and to avoid any contamination and pollution with the environment. If used, softeners (plasticisers) are usually added with the fillers to aid their dispersion. To increase both the dispersive degree and the

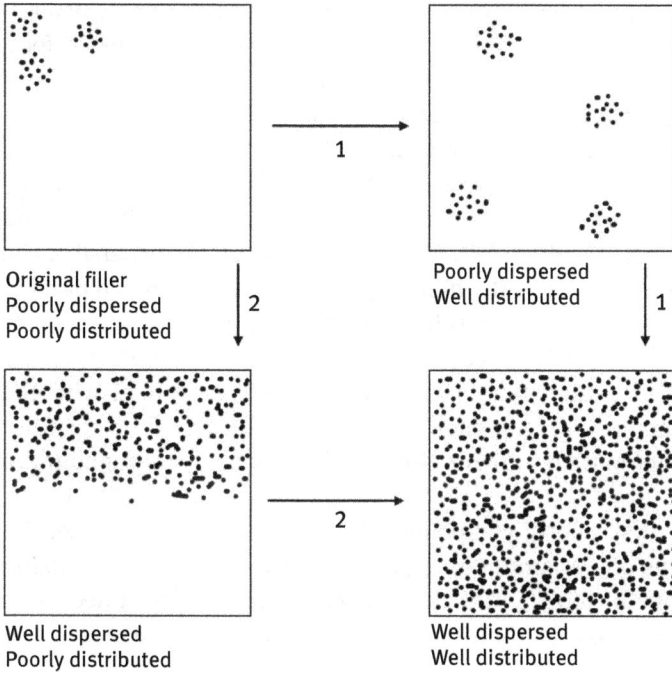

Fig. 2.2: Scheme of routes for reaching the best dispersion and distribution during compounding in the blending phase.

homogeneity of the system, processing aids can be added as well. Both these additives minimise the energy state of the process.

During blending, the filler agglomerates are smashed into smaller entities as a consequence of the high shear stress created in the masticator chamber. Figure 2.3 illustrates the transition from large agglomerates into smaller aggregates and primary particles, giving an indication of their dimensions. Agglomerates are continuously broken and aggregates with an average size of 100 nm appear. Smaller aggregates

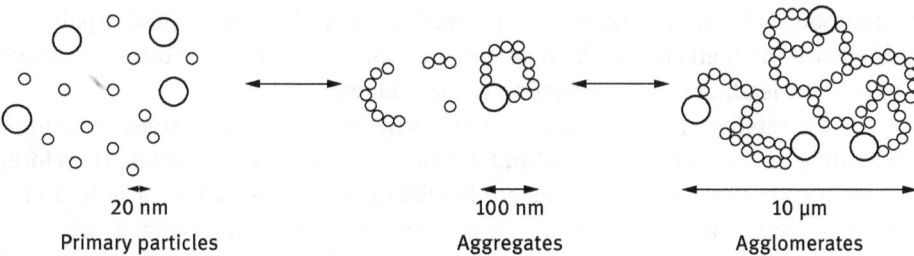

Fig. 2.3: Scheme of the transition from agglomerates to primary particles for a given filler.

and primary particles appear on the expense of larger aggregates and agglomerates. An optimal dispersion distributes fillers throughout the polymeric matrix in the form of aggregates. A poorer dispersion results in larger agglomerates. However, there is a lower limit to the aggregate size as the properties deteriorate with very small aggregate sizes and an increased number of primary particles.

During blending, the system is subjected to stretching deformation by shearing forces, which increases the interface between the disperse phase (fillers) and the elastomeric matrix, thus resulting in a gradual insertion of the disperse phase into the matrix itself. More precisely, the blending process consists in two parallel mechanisms: the filler wetting by the rubber matrix and the macromolecules deformation as a consequence of bonds tearing on adjacent chains. Bonds tearing allows the rapid coating of agglomerates by the polymer. In parallel with the reduction of the agglomerate size, the interface between the matrix and the filler is increased, and the filler particles are distributed homogenously throughout the rubber matrix.

After blending, the first step of compounding is concluded. On average, the whole mixing time of this phase is rather long and the temperature during mixing increases to reach a value of 160 °C. The global viscosity of the system decreases during mixing due to the opposite contributions of the temperature increase and the polymer breakdown.

The second step of compound preparation is characterised by the addition of the vulcanising agents and accelerators. The needed time for reaching the adequate dispersion and homogeneity is shorter than that for the first step of compounding. In addition, the maximum temperature is lower (max 120 °C) owing to the possible reaction of curatives with the polymer. The decrease of viscosity is not so high as in the first step. In any case, besides the vulcanising system, it is also convenient to add some vulcanisation inhibitors and retarders.

2.5 Compounding tools

As explained, the main goal of compounding is to provide the best homogeneity of the mixture, reaching the most uniform distribution of chemicals in the whole rubber compound [1, 4, 5]. Mixing is performed in specialised machines capable of dealing with the high stresses involved in shearing rubber, such as internal mixers or open mills leading to the production of material batches.

A batch mixing process typically consists of three sequential steps: weighing and loading of ingredients, mixing and discharge of the mixed product. Therefore, in a batch operation, all ingredients are loaded into the mixer together or in a predefined sequence, and mixed until a homogenous material is produced, which is discharged from the mixer in a single lot.

An open mill consists of twin horizontal counter-rotating rolls, one serrated, that provide an additional mechanical working to the rubber. The rolls can be heated or cooled as necessary. These rolls turn towards each other with a pre-set adjustable gap to allow the rubber to pass through to achieve the high-shear mixing. Mixing is achieved by the shearing action induced at the 'nip' between the rolls when the ingredients are incorporated in carefully weighed quantities. The back roll usually turns at a faster surface speed than the front roll. This difference in the rolls speed, called as friction ratio, increases the shear forces.

Generally, rubber forms a *bank*, called also *band*, around the front roll. The operator mixes the compound by cutting it off the rolls and refeeding it into the nip until all the ingredients are added and well incorporated in the elastomeric matrix as well. After the incorporation of all the non-curative ingredients, curing agents and accelerators are added. When the mixing operation is completed, the compound is removed from the mill in the form of sheets. The total mixing time is around 30 min for a batch size of approximately 30 kg in the biggest mill mix.

Mill mixing is the oldest and simplest method of rubber mixing, dating back to the very beginning of the rubber industry. Due to the long time needed for the filler incorporation, the limited batch size, and the poor ingredient dispersion achieved, the open mill has been replaced with other tools as the internal mixer. Currently, mill mixing is mainly used as a second-stage mixing device in the rubber compounding cycle, for adding the vulcanising agents and for completing the ingredient dispersion. Again, it is generally used for laboratory and small volume production.

Internal mixers are high shear machines developed by F. H. Banbury in 1916. Currently, the name 'Banbury' is a trademark owned by the Farrel Corporation. Today, they are commonly used because they are much more productive than the two roll mills, allowing a better filler dispersion and reducing the mixing time. Sizes of internal mixer from 40 to over 400 kg per batch are used in different areas of the rubber industry.

In the Banbury mixer, two slightly spiralled rotors revolve side by side toward each other within an enclosed chamber shaped like two short cylinders lying together with adjacent sides open (Fig. 2.4). These intersect to leave a ridge between the blades. At the top of the mixing chamber, there is a plunger, or hopper (ram), which can be lowered and raised between the open and closed positions by pneumatic or hydraulic action for allowing the Banbury charge.

The two counter-rotating rotors can be either tangential (non-intermeshing) or intermeshing. There are also a number of variations of these two, but in general the types with intermeshing rotors are preferred for heat-sensitive compounds. The rubber breakdown is accomplished by the shearing action between the rotor blades and the chamber walls and between the two rotors, with one running slightly faster than the other one to prevent the material sticking to the blades. This shearing

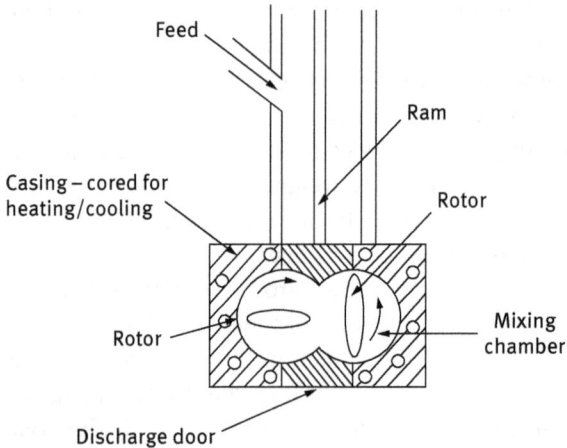

Fig. 2.4: Internal mixer with an intermeshing arrangement of rotors.

action generates considerable heat, so both the rotors and the chamber are water cooled to maintain the temperature low enough to assure that vulcanisation does not begin.

Mixing can be carried out in three or four stages, just to incorporate all the compound ingredients in the desired order. Ingredients are added through the ram at the top or through separate pipes (i.e., carbon black, mineral fillers and plasticisers) directly into the mixing chamber in roughly the same order as when mixing on an open mill. The number of revolutions per minute of the rotors can be varied even while the mixing cycle is in progress for improving the process efficiency. For example, it may be increased when plasticisers are incorporated since the viscosity and the degree of dispersion would otherwise normally decrease. In this way the time to mix a batch is reduced to under the typical mixing time of 5 min. The mixing time is also determined by the shape and the size of the rotors, the rotor speed and the power of the whole machine.

Upon completion of the mixing cycle, the rubber batch is rapidly discharged through the drop door at the bottom of the mixer. The compound needs to be quickly sheeted out and cooled to avoid scorch. It may be dumped into an extruder to form a continue sheet on a two-roll mill with an overhead mixer (stockblender) to improve the final homogenisation of the compound and to form a sheet. The latter is transported by a conveyor equipment to the cooling section. This operation takes several minutes, allowing some cold air to be blown over the rubber sheets. In the meantime an anti-tack agent is sprayed on both sides of the sheets. After this procedure, the cooled rubber can be placed on pallets for further transportation.

References

[1]   Dick, JS. Rubber technology – Compounding and testing for performance. Hanser, Munich, Germany, 2001.
[2]   Rodgers, B. Rubber compounding. Marcel Dekker, New York, USA, 2004.
[3]   Hamed, GR. Engineering with rubber. Hanser, Munich, Germany, 2001.
[4]   Kim PS, White Jl. Simulation of Flow in an Intermeshing Internal Mixer and Comparison of Rotor Designs. Rubber Chem. Tech. 1996, 69, 686.
[5]   White, JL. Development of Internal-Mixer Technology for the Rubber Industry. Rubb. Chem. Tech. 1992, 65, 527.

3 Vulcanisation

3.1 The vulcanisation process

As already introduced in the former chapters, during vulcanisation the flexible rubber chains are joined together by cross-linking reactions, giving a three-dimensional network with peculiar physical–mechanical properties [1]. Thus during vulcanisation, rubber is converted from a plastic, soft, sticky substance having very low strength and elongation at break to a resilient highly elastic material of considerable strength. Since the quantity (degree) and quality (type) of cross-links determine the properties of the resulting network and those of the end-products, an understanding of the formation, structure and stability of vulcanisates is essential.

Vulcanisation occurs between two statistically favourable reactive sites, such as the double bonds in diene-containing elastomers or the double bonds pendant to the polymer backbone from cure-site monomers. The reactive sites can also be formed by the abstraction of backbone hydrogen or halogen atoms in the main elastomeric chains. The reactive sites are covalent bonded through one of the following mechanisms:

- Insertion of difunctional curatives, such as sulphur, between the reactive sites.
- Carbon–carbon bond formation between the backbone chains by a free-radical mechanism initiated by a peroxide vulcanising agent or by high-energy radiation.
- Insertion of di- and multifunctional monomers such as acrylates, phenolics or triazines between the reactive sites.

Since different types of vulcanisation systems are available, the choice of the optimal system for a given application depends on the required curing conditions, the elastomer and the expected physical properties in the final vulcanisates. For example, the most widely used elastomers containing a diene site (i.e., natural rubber, polyisoprene, butadiene rubber, styrene–butadiene rubber, butyl rubber and nitrile–butadiene rubber) are all cross-linked using sulphur as the vulcanising agent.

Any process of vulcanisation is successful only if it can be controlled. It should begin when required, accelerate when needed and stop at the right time. In the jargon of rubber technologists, these steps are termed as scorch resistance, acceleration and cure time, respectively. Scorch resistance is the time elapsed before the vulcanisation starts. In this induction period, the vulcanising agents slowly react with the rubber and the other additives; thus, soluble rubber intermediate products are formed. When the scorch time is too short, premature vulcanisation can occur, resulting in the development of cracks and making unusable the resulting vulcanisates. Once vulcanisation begins, it should be completed as fast as possible to have

https://doi.org/10.1515/9783110640328-003

a reasonable production cycle. Indeed, for economic reasons, shorter cure times are preferred.

With a complete understanding of the relationship between vulcanisation chemistry and network structure, rubber formulations to produce the desired mechanical and chemical properties can be properly tailored.

3.2 Influence of cross-link type and density on the rubber behaviour

Both the nature and the number of cross-links formed, also referred to as the degree of vulcanisation, have an influence on the final properties of vulcanisates [2]. In particular, the presence of the cross-linked network determines the following:

1. Insolubility of the cross-linked elastomer: vulcanisates can only swell.
2. Strength increases till the value determined by the curing process. In the case of over-cure with respect to the optimal stage of vulcanisation, the strength decreases, but both the modulus and the hardness increase. When natural rubber over-cures, both the modulus and the tensile strength decrease because of the phenomenon of reversion.
3. Improvement of both strain and dynamic fatigue resistance due to the increasing of the vulcanisation degree.
4. Low sensitivity of vulcanisates due to temperature variations.
5. Elasticity of vulcanisates over a wide temperature region.

The aforementioned points highlight how cross-links influence the mechanical properties of vulcanised elastomers. To better understand these effects, the cross-link density, the cross-link type and other parameters related to the vulcanisation process should be taken into account.

3.2.1 Average molecular weight

The average molecular weight (M_c) is defined as the average mass of a polymer chain that connects two adjacent cross-links. The practical M_c value is typically between 8,000 and 10,000 (114–142 CRUs - constitutional repeating unit), considering the average molecular weight of the individual monomeric units (CRU) to be 70.

M_c is related to the tensile strength of the vulcanisates, as shown in Fig. 3.1, by plotting the shear modulus versus temperature. When raw rubber is stressed, the macromolecular chains readily slide over one another and detangle. The introduction of few cross-links increases the molecular weight of the chain, creating a branched molecule and a broader molecular weight distribution. In this way, the

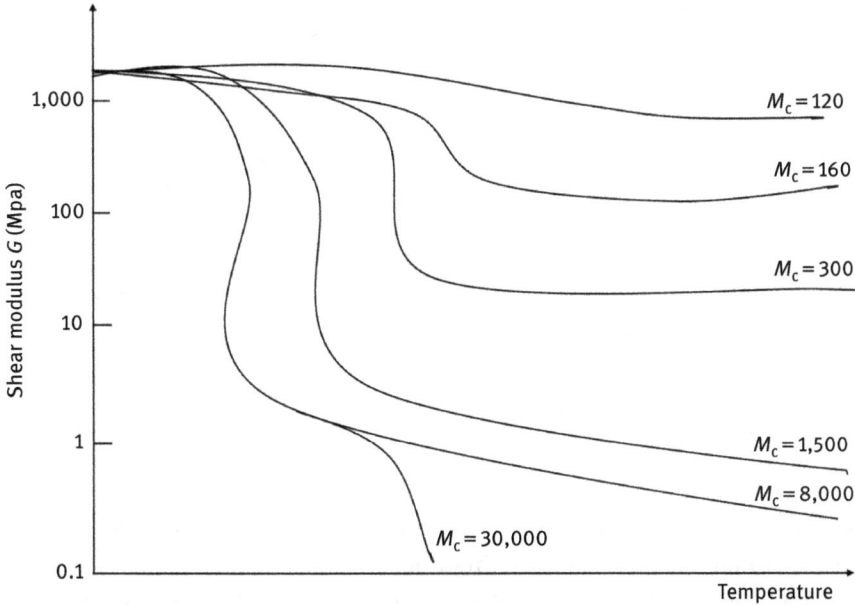

Fig. 3.1: Influence of M_c (average molecular weight between two adjacent cross-links) on the shear modulus behaviour as a function of temperature.

branched chains are hampered to detangle, thus increasing the tensile strength of rubber. Again, high molecular weight of segments between cross-links should allow a greater number of polymer chains to become load bearing under strain, resulting in higher apparent tensile strength.

3.2.2 Type and degree of cross-linking

The degree and type of cross-links are very important factors for achieving the required vulcanisate properties, as described later in the case of sulphur-based vulcanisation. The degree of cross-linking is expressed as the moles of cross-linked basic units per total moles of basic units.

In the vulcanisation network, sulphur may be combined in a number of ways. It can remain as monosulphide (S), disulphide (S_2) or polysulphide (S_x) (cross-links consisting of more than two sulphur atoms). It may also be present as pendent sulphides, cyclic monosulphides or disulphides (intramolecular cycles), as shown in Fig. 3.2.

The type of developed cross-links depends on sulphur dosage, accelerator type, accelerator/sulphur ratio and cure time. The monosulphidic cross-links are created by using sulphur donors (i.e., thiuram disulphide or morpholine

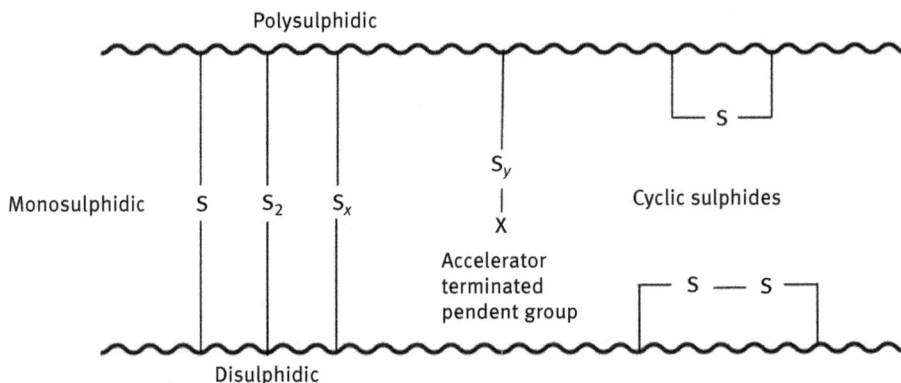

Fig. 3.2: Vulcanisation network formed by sulphur.

disulphide). The di- and polysulphidic cross-links are formed by varying the accelerator/sulphur ratios. Generally, high accelerator/sulphur ratio and longer cure time increase the number of monosulphide cross-links at the expense of the polysulphide ones. Such vulcanisates exhibit better heat stability, lower compression set and longer reversion time when compared to polysulphide predominant network, due to the better stability of C–S bonds with respect to S–S bonds. On the other hand, vulcanisates containing higher amounts of polysulphide cross-links offer higher tensile strength, tear strength and flex-fatigue resistance due to the ability of S–S bonds to break reversibly and to locally release the high stresses, which could lead to failure.

Cross-links that form bridges between the chains are stress-bearing members contributing to elasticity and strength of rubber. Cyclic sulphides, accelerator fragments and vicinal cross-links do not contribute to elasticity.

3.2.3 Cross-link density

As schematically shown in Fig. 3.3, the vulcanisate properties strongly depend on the cross-link density, defined as the moles of cross-linked basic units per weight unit of the cross-linked polymer [3,4]. Hysteresis generally decreases with an increase in the cross-link density. Fracture properties, such as tensile strength and tear strength, reach a maximum value as the cross-link density increases and then decrease. Again, increasing the cross-link density, elongation and permanent set decrease. Properties such as static modulus, dynamic modulus and hardness increase with an increase in the cross-link density. The static modulus increases with vulcanisation to a greater extent than the dynamic modulus. The dynamic modulus is a composite of viscous and elastic

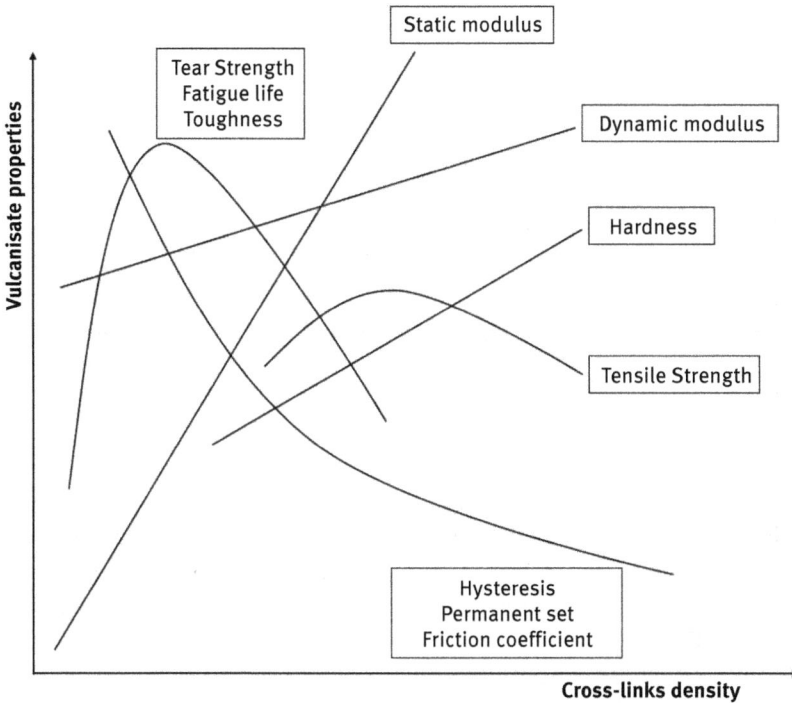

Fig. 3.3: Influence of cross-links density on the vulcanisate's properties.

responses, whereas the static modulus is a measure of the elastic component alone. Vulcanisation, then, causes a shift from the viscous or plastic behaviour to elasticity.

Cross-link density has to be high enough to prevent the fracture due to the viscous flow but low enough to avoid brittleness. In addition, the cross-link mobility (ability to realign under strain) should further enhance the tensile strength.

Generally, the mechanical behaviour of any rubber can be related to the chemical nature of cross-links:

- S_x *polysulphidic*: The maximum tensile strength is reached with one cross-link every 150 monomer units. The extra sulphur is available to create additional cross-links, which raise the modulus of the compound.
- *S monosulphidic*: Rubber shows about 30% lower tensile strength at a slightly lower cross-link density. This could be due to the minimal ability of the mono-sulphidic cross-link to distribute the mechanical strain.
- *C–C cross-links, peroxide initiated*: The material has a tensile strength about 40% lower than that of the S_x network. Again, this is probably due to the immobility of the C–C bond.

3.3 Vulcanising agents

Several different vulcanising agents (curatives) are currently used in rubber industry [1]. Together with the most common sulphur, organic peroxides, quinones, metal oxides, bifunctional oligomers, resins and amine derivatives are used as well. Vulcanisation can also be achieved by using high-energy radiation without some chemicals.

Any vulcanisation reaction is determined in large measure by the type of vulcanising agents, the temperature and the curing time. Any vulcanising system should give not only a rapid and effective cross-linking at the desired vulcanising temperatures but also should resist to the premature scorching at lower temperatures that may be required to mix, extrude, calender or subject the rubber and to other shaping operation before cross-linking.

Sulphur is the original and still the most widely used curative because of its versatility and cost. It exists in the elemental state as an eight-membered ring (S8). The sulphur ring-opening mechanism, as discussed by Coran, involves either a free radical or an ionic mechanism [1]. The ionic mechanism is probably more logical and can be rationalised in terms of the generalized Lewis acid–base interactions, as discussed by Jensen [5].

The basic ingredients for a sulphur cure are as follows: sulphur or sulphur donor (cross-linker), organic accelerator(s), zinc oxide and fatty acid (activator). Vulcanisation of rubber with sulphur alone (without any accelerator) is a very slow process, and it takes several hours or even days to reach the optimum cure depending on the vulcanisation temperature and the nature of rubber. The resulting vulcanisates show very poor mechanical and ageing resistance, with a strong tendency for reversion. Vulcanisation with sulphur alone is, therefore, of no technological importance at all. A major breakthrough came with the discovery of organic nitrogen compounds used as accelerators. Bases like aniline and thiocarbanilide were the first vulcanisation accelerators of rubber. Now a large number of compounds are used that reduce the curing time and the oxidative degradation of the elastomers during vulcanisation, thus improving the end properties of vulcanisates.

Saturated elastomers, such as ethylene propylene rubber, silicone, fluoroelastomers and so on are cross-linked by peroxides. Networks formed from peroxide curing typically possess good heat-ageing stability and low compression set since cross-links are very stable covalent C–C bonds. The synergistic use of multifunctional coagents can improve the vulcanisate properties by increasing the cross-link density of the network and by altering the cross-link composition. Since the final properties of the formed network will depend on the reactivity and structure of the coagent, the understanding of these structure–property relationships will allow a better coagent selection among the available functional compounds.

Sulphur and non-sulphur systems have both advantages and disadvantages, but sulphur systems still remain the most versatile ones because of the following benefits: higher flexibility during compounding, easier adjustment of the balance between the vulcanisation stages, possibility to control the cross-links length, better mechanical properties of vulcanisates and economical reasons. However, compared to peroxide curing, sulphur systems show lower heat and reversion resistance, higher compression set and higher possibility of corrosion in cable metals.

3.4 Accelerators

Accelerators are added in small amounts during compounding to accelerate the vulcanisation reaction and to improve the physical and service properties of vulcanisates [6]. The shortening of the vulcanisation time is generally accomplished by changing the amounts and/or the types of accelerators used. The decrease in the vulcanisation time is of tremendous economic importance because it allows to reduce the production cost. Furthermore, the amount of the required vulcanising agent can be considerably decreased in the presence of an accelerator. Almost all accelerators need activators for the development of their full activity: zinc oxide is still being used as the best additive (see Section 3.5).

Accelerators can be classified following different criteria, such as their influenceon the vulcanisation rate, thus distinguishing slow-, high- and medium-activity accelerators and ultra-accelerators. However, a classification based on the chemical composition of accelerators seems to be rational and currently it is the most used one. Tab. 3.1 lists the most important types of organic accelerators classified based on the chemical families to which they belong.

Accelerators can also be classified as primary and secondary based on the role they play in a given compound. When the accelerator is present at a relatively higher concentration (from 0.5 to 1.5 phr) it is called a primary accelerator, whereas that in smaller amounts it is called as a secondary accelerator. The dosages of the secondary accelerator are generally between 10% and 40% of the primary accelerator.

A binary accelerator system refers to the use of two or more accelerators in a given formulation [7]. Binary systems are widely used in industry since they provide better acceleration, better control of processing safety and an improvement in the physical and chemical characteristics of final vulcanisates, thanks to the synergistic behaviour of the two accelerators [8]. Generally, thiazoles and sulphenamides play the role of primary accelerators due to their intrinsic characteristics, such as a good processing safety, the broad vulcanisation plateau and the optimum cross-link density as well as the required reversion delay that they offer. Basic accelerators such as guanidines, thiurams and dithiocarbamates are used as secondary accelerators to activate the primary ones. The most common binary systems consist of thiocarbamate derivatives and benzothiazoles. Diphenyl guanidine is also used in

Tab. 3.1: Classification of organic accelerators based on their chemical compositions.

Type and formula	Chemical name	Properties
Aldehyde amines		
	Hexamethylene tetramine (hexamine) (HMT)	Occasionally used as a secondary accelerator
$CH_3CH=NH-$	Ethylidene aniline (EA)	Medium accelerator, used mainly with thiuram sulphides and dithiocarbamates
Guanidines	Diphenyl guanidine (DPG)	Medium accelerator, used mainly with other accelerators
	Triphenyl guanidine (TPG)	Slow accelerator
	Di-o-tolyl guanidine (DOTG)	Medium accelerator, also used as a plasticiser for neoprene. It disperses more readily than DPG
Thiazoles	Mercaptobenzothiazole (MBT)	Semi-ultra-accelerator, apt to be scorchy
	Dibenzothiazyl disulphide (MBTS)	Semi-ultra-accelerator with a delayed action
	Sodium salt of MBT (NaMBT)	Water soluble, used in latex compounding

Tab. 3.1 (continued)

Type and formula	Chemical name	Properties
	Zinc mercaptobenzothiazole (ZMBT)	Semi-ultra-accelerator
	2,4-Dinitrophenyl mercapto-benzothiazole (DMB)	Semi-ultra-accelerator
Sulphenamides 	N-cyclohexylbenzo-thiazolsulphenamide (CBS)	Semi-ultra-accelerator with a delayed action
	N-oxydiethylbenzo-thiazolsulphenamide (NOBS)	Semi-ultra-accelerator with a delayed action
	N-t-butylbenzothiazol-sulphenamide (TBBS)	Semi-ultra-accelerator
	N,N'-dicyclo-hexylbenzothiazol-sulphenamide (DCBS)	Semi-ultra-accelerator
Dithiocarbamates 	Piperidine pentamethylene dithiocarbamate (PPD)	Ultra-accelerator
	Zinc diethyl dithiocarbamate (ZDC, ZDEC)	Ultra-accelerator
	Sodium diethyl dithiocarbamate (SDC)	Water-soluble ultra-accelerator, used for latex
	Zinc ethyl phenyl dithiocarbamate	Scorch-resistant ultra-accelerator

(continued)

Tab. 3.1 (continued)

Type and formula	Chemical name	Properties
Thiuram sulphides	Tetramethyl thiuram disulphide (TMT, TMTD)	Ultra-accelerator
	Tetraethyl thiuram disulphide (TET, TETD)	Ultra-accelerator
	Tetramethyl thiuram monosulphide (TMTM)	Ultra-accelerator
	Dipentamethylene thiuram tetrasulphate (DPTS)	Ultra-accelerator with tendency to scorch
Xanthates	Zinc isopropyl xanthate (ZIX)	Ultra-accelerator
	Sodium isopropyl xanthate (SIX)	Water-soluble ultra-accelerator for latex
	Zinc butyl xanthate (ZBX)	Low-temperature ultra-accelerator

combination with MBT or sulphenamides to increase the scorch delay period. Thiuram systems generally show very little scorch safety.

Although the practice of using binary accelerators is quite old, the mechanism of the combined action of these accelerators has not been studied adequately and only recently scientists began to fully probe the complicated mechanism of binary systems. The higher accelerating action of binary systems is assumed to be because of the formation of either an eutectic mixture or a salt compound showing greater chemical reactivity and better solubility with respect to having only one accelerator. Dogadkin and co-workers investigated a number of popular accelerator combinations, founding a mutual activation with many of them and in other cases a synergic behaviour [9–11].

3.4.1 Accelerators selection in rubber compounding

Before selecting the most suitable accelerator system for manufacturing a particular rubber compound, the following points should be taken into account [6]:
- solubility in rubber (high solubility is required to avoid bloom and to improve the dispersion);
- processing operations and temperatures that the rubber compound is to be required to undergo;
- adequate scorch time needed for having a scorch-free processing and a good storage stability;
- required reversion characteristics (delayed reversion on over-cure);
- vulcanisation method to be used;
- cure cycle, depending on the vulcanisation method and temperature;
- required vulcanisate's properties;
- no adverse effects on other characteristics or materials (i.e., bonding, ageing, adhesion, etc.)
- no health hazards due to its decomposition products and
- no adverse effects during the end-use of the rubber items, especially in the case of food contact and medical applications.

Thiazole accelerators

Thiazoles are medium-fast primary accelerators with only moderate processing safety. This can be further boosted to increase the vulcanisation rate by using small quantities of basic accelerators such as DPG, DOTG, TMTM, TMTD, ZDC and so on. They can be retarded using small proportion of phthalic anhydride, salicylic acid and so on. In addition, they can act also as retarders in rubber compounds accelerated with thiurams/dithiocarbamates systems.

Thiazoles are activated by zinc oxide in combination with stearic acid and produce flat cure with vulcanisates having very good reversion resistance. Generally, they display a delayed action and give vulcanisates with good elastic properties, such as cycle tyres and tubes, footwear, beltings, hoses and other moulded and extruded goods. Thiazoles are particularly preferred in rubber for metal-bonding applications, since amine accelerators affect the rubber–metal bond strength.

The activity of thiazole accelerators with respect to the cure characteristics can be summarised as follows:
1. Scorch safety → longer (ZMBT < MBTS < MBT)

2. Cure rate → faster (ZMBT < MBTS < MBT)
3. Cross-link density → higher at equal dosage (ZMBT < MBT < MBTS)

Sulphenamide accelerators

Sulphenamide accelerators are the most popular in the tyre industry due to their delayed action as well as the faster cure rate offered in the vulcanisation of rubber compounds containing furnace blacks.

At the beginning, the development of sulphenamide accelerators was focused on improving the scorch safety of NR-based tyre compounds by considering their storage stability. New sulphenamide accelerators have been introduced later to accommodate the extra scorch safety required for processing the fine particle-sized hard carbon blacks, which offer higher abrasion resistance for tyre-tread compounds.

Sulphenamides provide a wide range of cross-link densities depending on their type and dosage and display flat and reversion-resistant cure. Progressive increase in their amounts leads to an improvement in scorch delay, cure rate and state of cure. Vulcanisates of sulphenamide accelerators have a typical 'aminic' odour and exhibit higher stress–strain properties along with better resilience and flex–fatigue resistance when compared to thiazoles. These accelerators have a limited storage stability and the rate of degradation is greatly influenced by the storage conditions, such as humidity and heat. They are usually incorporated in the rubber compound at the end of mixing cycle when the temperature is above their melting point to ensure a proper dispersion. Generation of excess heat is avoided to prevent their decomposition.

The activity of sulphenamide accelerators can be summarized as follows:
1. Scorch safety → longer (CBS < TBBS < DCBS)
2. Cure rate → faster (DCBS < CBS < TBBS)
3. Cross-link → higher at equal dosage (DCBS < CBS < TBBS)

Thiuram accelerators

Thiuram sulphides are used as both primary or secondary accelerators. Vulcanisates have good ageing and heat resistance and are tasteless. Thiurams do not stain and/or discolour the final products.

They are ultrafast accelerators for NR, SBR, BR, NBR and other highly unsaturated rubbers and the most preferred primary accelerator for sulphur cured low-unsaturation content rubbers such as butyl and EPDM. Thiurams are also widely used as a secondary accelerator with thiazole/sulphenamide systems to achieve faster curing rate and higher cross-link density with a good compromise on scorch safety. Thiurams are also employed along with guanidine in polychloroprene compounds to attain good processing safety. In combination with dithiocarbamates and xanthates, they display a retarding effect without changing the rate of vulcanisation.

The activity of various thiuram accelerators with respect to the cure characteristics can be summarised as follows:
1. Scorch safety → longer (TMTD < TETD << TMTM)
2. Cure rate → faster (TMTM = TETD = TMTD)
3. Cross-link → higher at equal dosage (TMTM = TETD = TMTD)

Dithiocarbamate accelerators
Dithiocarbamate accelerators are widely used as ultra-fast accelerator for NR latex-based compounds and also find applications as primary or secondary accelerators in most sulphur-based rubbers. They require zinc oxide and stearic acid for the activation and give faster vulcanisation. They exhibit very low scorch safety, faster cure rate and higher cross-link density. Compounds accelerated with dithiocarbamates have a very narrow plateau; hence, reversion due to over-cure can take place very rapidly.

Dithiocarbamates are non-staining and non-discolouring even on exposure to light and are suitable for the manufacture of transparent goods. Vulcanisates show high breaking strength and low compression set. However, they show a limited solubility in rubber compounds and hence excess quantity tends to bloom onto the surface of vulcanisates.

The activity of various dithiocarbamate accelerators with respect to the cure characteristics in rubber compounds can be summarised as follows:
1. Scorch safety → longer (ZDMC < ZDEC < ZDBC)
2. Cure rate → faster (ZDBC = ZDEC = ZDMC)
3. Cross-link → higher at equal dosage (ZDBC = ZDEC = ZDMC)

Aldehyde amines
Currently, aldehyde amines have a limited use, exclusively as secondary accelerators, due to their high toxicity and the staining effect. They are active at high temperature and produce vulcanisates with good ageing resistance. Their activity is inhibited by the acidic compounds that exist in rubber like tars, channel black and so on.

Aryl guanidines
They are secondary accelerators, used often with thiazole accelerators to produce vulcanisates with low ageing resistance and a slight tendency to a yellow shade.

Xanthates
These accelerators are sodium and zinc salts of alkyl xanthic acid, used for the vulcanisation of latexes or for self-vulcanising solutions.

3.5 Activators

As already mentioned, activators are molecules able to increase the efficiency of the accelerators [7]. In the presence of the accelerator/activator pair, there is an increase of vulcanisation rate, a reduction of vulcanisation temperature and time and an increase of the mechanical properties of vulcanisates. The action mechanism of activators is complex and it is specific to each elastomer–accelerator–activator system.

The most common inorganic compounds used as activators for rubber are:
- Zinc oxide (ZnO): the most known vulcanising activator, used along with the majority of accelerators. Zinc oxide has been an important ingredient since the early days of rubber compounding. Originally used as an extender to reduce cost, it was subsequently found to have a reinforcing effect and was later found to reduce the vulcanisation time.
- Magnesium oxide (MgO): mainly used for the vulcanisation of the neoprene-type elastomers.
- Litharge (PbO) and minimum (Pb_3O_4): less used, but of interest in the case of systems containing thiazole, dithiocarbamate or thiuram sulphide accelerators.

Usually a combination of zinc oxide and a long-chain fatty acid (aliphatic carboxylic acids), such as stearic acid, is the most used activator system, in which the zinc ion is made soluble by salt formation between the acid and the oxide. Fatty acids, besides the activation effect, improve the dispersion of ZnO in the elastomeric matrix.

3.6 Scorch retarders

A proper balance of rubber-processing safety and faster curing rates is essential for increasing both productivity and the economic use of high-value rubber-processing equipment [6]. Often higher processing temperatures are used along with faster accelerator combinations; in addition, higher vulcanisation temperatures are employed to reduce the vulcanisation time. Sometimes rubber-processing temperatures can induce the premature vulcanisation of the compound, known as scorch. The scorch at an advanced stage makes the rubber compound useless for further processing or vulcanisation: only just few scorched particles can drastically reduce the physical properties of vulcanisates.

To avoid scorch with faster curing systems, higher processing temperatures and prolonged storage periods, scorch retarders or pre-vulcanisation inhibitors can be added to the compound. More precisely, these ingredients can reduce the accelerator activity during both processing (mixing, preheating, extrusion and calendering) and storage since they extend the scorching period without decreasing the vulcanisation rate. Clearly, they should either decompose or not interfere with the accelerator system during the normal curing at elevated temperature.

Retarders are organic acids, such as salicylic acid and phthalic anhydride, which act by lowering the pH of the compound, thus retarding vulcanisation. Currently, *N*-cyclohexylthiophthalimide is the largest retarder used in the rubber industry. *N*-nitrosodiphenylamine and thiosulphenamides also constitute special class of retarders. Generally, scorch retarders display the maximum activity in relation to certain vulcanising accelerators. Thus, *N*-nitrosodiphenylamine is more active in the presence of thiazole, thiuram and dithiocarbamate accelerators, while it is ineffective with aldehyde amines.

3.7 Accelerated sulphur vulcanisation

The most common vulcanisation systems used in industrial applications are the accelerated sulphur formulations [6, 7]. A typical sulphur vulcanisation system comprises zinc oxide (3–10 phr), stearic acid (1–4 phr), accelerator (0.5–4 phr) and sulphur (0.5–3 phr). There are also accelerator systems as tetra-methylthiuram disulphide, in which the elemental sulphur is not present: the accelerator itself provides sulphur for vulcanisation. This sulphur-free vulcanisation can also be referred to as sulphur donor systems.

To understand the mechanism of rubber vulcanisation with sulphur in the presence of accelerators and activators, many studies have been carried out, and currently, the majority of the involved processes are known. Since a series of consecutive and competing reactions occur during curing, single mechanism cannot be appropriate. Both radical and ionic reactions are involved during sulphur vulcanisation and the resulting effect largely depends on the compounding formulation. These reactions comprise double bond migration, isomerisation, chain cleavage, cyclisation and formation of vicinal cross-links [12–17].

A generally accepted scheme of the vulcanisation mechanism with sulphur is as follows [14, 18]:

1. Accelerator (Ac) and activator interact with sulphur to form the active sulphurating agent:

    $Ac + S_8 \rightarrow Ac–S_x –Ac$ (*active sulphurating agent*)

2. The rubber chains interact with the sulphurating agent to form polysulphidic pendant groups terminated by accelerator groups:

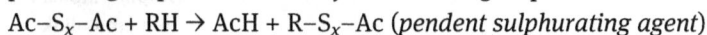

    $Ac–S_x–Ac + RH \rightarrow AcH + R–S_x–Ac$ (*pendent sulphurating agent*)

    where RH is the rubber chain.

    In this pendant group, a fragment derived from the accelerator or the sulphur donor is linked through two or more sulphur atoms to the rubber chains. Cross-links are formed either by direct reaction with another rubber molecule or by disproportionation with a second pendent group of a neighbouring rubber chain.

3. Polysulphidic cross-links are formed:

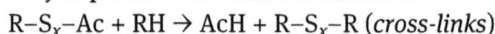

    $R–S_x–Ac + RH \rightarrow AcH + R–S_x–R$ (*cross-links*)

4. The network is formed, thanks to a number of competing side reactions [7]:

    $R-S_x-Ac \rightarrow$ cyclic sulphides + dienes + ZnS (*degradation*)

    $R-S_x-Ac \rightarrow S_{x-1} + R-S-Ac$ (*desulphuration*)

    $R-S_y-R \rightarrow S_{y-1} + R-S-R$ (*monosulphidic cross-links*)

    $R-S_{x+y}-R + AC-S_2-Ac \rightarrow AC-S_{y+2} -Ac + R-S_x-R$ (*sulphur exchange*)

    Polysulphide cross-links undergo a further transformation by two competing reactions, that is, desulphuration or degradation. The progressive shortening of the polysulphide produces finally monosulphide links by desulphuration.

Considering the aforementioned mechanism, the resulting cross-linked network depends on the nature of rubber, the ratio sulphur/accelerator and the vulcanisation temperature. In particular, the accelerator/sulphur ratio determines the efficiency by which sulphur is converted into cross-links, the nature of cross-links and the extent of the main chain modification.

The sulphur vulcanising systems for highly unsaturated general-purpose rubbers can be categorised on the basis of the sulphur/accelerator ratio:

- conventional or high sulphur vulcanisation system (CV), where sulphur is added in the range of 2–3.5 phr and the accelerator in the range 1–0.4 phr;
- efficient vulcanising system (EV), where sulphur is added in the range of 0.3–0.8 phr and the accelerator in the range of 6.0–2.5 phr; and
- semi-efficient system (SEV), where sulphur is added in the range 1–1.8 phr and the accelerator in the range 2.5–1 phr.

The number of sulphur atoms in a cross-link depends on the type of the cure system chosen for the vulcanisation process, as listed in Tab. 3.2. As the CV system contains a larger amount of sulphur when compared to the accelerator, the possibility of polysulphidic linkage formation is higher. Its vulcanisates exhibit good tensile and tear strength, good fatigue and low temperature resistance. However, at higher temperatures, the polysulphidic links may break to mono- and disulphidic, thus explaining the reversion phenomenon, which leads to a decrease in both strength and modulus.

Tab. 3.2: Number of sulphur atoms per cross-link depending on the vulcanisation system.

Vulcanisation system	Sulphur atoms per cross-link
Un-accelerated sulphur vulcanisation	40–45
Conventional (accelerator-sulphur) cure (CV)	10–15
Semi-efficient vulcanisation (SEV)	5–10
Efficient vulcanisation (EV)	4–5
Elemental sulphur less vulcanisation (sulphur donor cure)	Less than 4

EV systems may contain of a sulphur donor instead of an elemental sulphur or a combination of low concentration of elemental sulphur (less than 0.5 phr) and high concentration of accelerators. The EV system produces a network containing thermally stable monosulphidic and disulphidic cross-links with lesser chain modifications. The short sulphur cross-links provide poor tensile and tear strength, poor flex-fatigue life and low abrasion resistance. However, EV cure systems offer good heat ageing, thermal stability and compression set resistance. These cure systems are generally used for rubber products with thick cross-section and products with static applications.

The SEV cure systems are an attempt to find a compromise between CV and EV cure. They have found particular application in NR where a compromise between heat ageing and fatigue life is often necessary.

3.8 Peroxide curing

Peroxide curing involves the use of organic peroxides (R–O–O–R′) that at elevated temperatures decompose to form highly reactive radicals, which chemically cross-link the macromolecular chains [1]. The basic chemistry of peroxide decomposition and the subsequent crosslink-forming reactions are well established for various unsaturated and saturated elastomer systems [19].

In the cross-linking mechanism of peroxide vulcanisation three major steps are involved:

1. Peroxide decomposition: Peroxide undergoes the homolytic cleavage of the O–O bond to form two alkoxy radicals as a result of an energy input (heat):

$$R - O - O - R \rightarrow 2R - O^*$$

where R is an alkoxy, alkyl or acyloxyl radical, depending on the type of peroxide (e.g., benzoyl peroxide gives benzoyloxy radicals and dicumyl peroxide gives cumyloxyl radicals). When vinyl groups are also present in the polymer, the vulcanisation reaction is quicker and the main attack is undoubtedly at these reactive sites. When symmetrical peroxides are used, two radicals of similar activity are formed, which can initiate the cross-linking reactions.

2. The highly reactive alkoxy radical abstracts a hydrogen atom from a polymer chain:

$$R - O^* + P - H \rightarrow P^* + ROH$$

The dehydrogenation of the polymer chain and the subsequent transfer of radicality to the hydrogen chain occur. In the case of unsaturated hydrocarbon elastomers, such as butadiene or isoprene, the next step is the abstraction of a hydrogen atom from an allylic position on the polymer molecule or the addition of the peroxide-derived radical to a double bond of the polymer molecule. For isoprene rubber, the abstraction route predominates over the radical addition; thus, two polymeric free radicals combine to give a cross-link.

3. Cross-linking: Two radicals on adjacent polymer chains couple to form the C–C bond:

$$2P^* \rightarrow P-P$$

The formation of a new C–C bond (cross-link) is an important feature of peroxide vulcanisation. Neither the peroxide nor the by-products are part of the cross-link; hence, the inherent polymer stability is retained after cross-linking.

The rate-determining step in the curing process is the homolytic cleavage of the peroxide molecule, which is completed in few seconds or minutes. This is a first-order reaction, where the rate is proportional to the concentration of the peroxide present at any time and is also controlled by the energy available for the homolytic cleavage. The rate of the radical formation also depends on the temperature of the system. Most peroxides that are commonly used in rubber formulations are very stable and require high temperatures (above the rubber-processing temperatures) for decomposition.

Different types of peroxides are used in rubber curing. Peroxides with carboxylic acid groups (i.e., diacetyl peroxide and dibenzoyl peroxide) exhibit low sensitivity to acid groups, have lower decomposition temperatures and are not useful in the presence of carbon black. Peroxides without carboxylic groups (i.e., di-tert-butyl peroxide and dicumyl peroxide) are very sensitive to acids, exhibit higher decomposition temperature and are less sensitive to oxygen. Aliphatic substituted peroxides (i.e., tert-butyl cumyl peroxide) are favourable over aromatic-substituted peroxides (i.e., tert-butyl perbenzoate).

Organic peroxides are useful for cross-linking saturated as well as unsaturated polymers or those that contain no sites available for the attack by other vulcanising agents. They are useful for ethylene–propylene rubber, silicone, HNBR and fluoroelastomers. They are not generally useful for vulcanising butyl rubber, poly(isobutylene-co-isoprene) and epichlorohydrin, because of their tendency towards chain scission rather than cross-linking when the polymer is subjected to the action of peroxides.

The main advantages of peroxide cure over sulphur cure are the simpler formulation, the long-term compound storage stability and the possibility of using higher processing temperatures. They rapidly cure at high temperature and show no reversion. Non-staining, non-blooming and non-discolouring effects are observed. However, peroxide cure have some drawbacks, such as the high sensitivity to oxygen during cure and to process oils, antioxidants, resins, acidic clays and other acidic materials used in compounding that can significantly affect the curing process and the vulcanisate properties (i.e., tensile strength, tear strength, flex-fatigue resistance, abrasion, etc.). Long cures and higher curing temperatures are required together with post-curing to remove the unwanted products and the foul odour. Generally, both peroxides and the whole vulcanisation process are more expensive than the sulphur systems.

Peroxides are co-activated by coagents that add to the system those unsatura-
tions that provides higher cross-link densities for a given peroxide concentration,
exhibiting also some plasticizing effect [1, 6, 7].

Coagents are classified into two basic classes based on their contributions to
cure: Type I and type II. Type I coagents are typically polar, low molecular weight
multifunctional compounds (i.e., maleimides and methacrylates) that propagate
very reactive radicals primarily through addition reactions, thus increasing both
the rate and the state of cure. Type II coagents include allyl-containing cyanurates,
isocyanurates, phthalates and high vinyl poly(butadiene) resins. They form less
reactive radicals and contribute only to the state of cure.

Because of their reactivity, coagents generally make a more efficient use
of the radicals derived from peroxides, whether acting to suppress non-net-
work forming side reactions during cure or to generate additional cross-links
[20, 21]. The mechanism of cross-link formation using coagents appears to be
at least partially dependent on their class. Most type I coagents exclusively
homopolymerise and form viable cross-links through radical addition reactions. Certain type II coagents, containing extractable allylic hydrogens, have
been shown to participate in intramolecular cyclisation reactions as well as
in intermolecular propagation reactions.

The reactivity of coagents can have negative effects on process safety, typically
manifested by a decrease in the scorch time. Type I coagents not only detract from
scorch safety but also provide faster cure rates, while type II coagents exhibit equivalent scorch safety, but have longer cure times.

3.9 Others vulcanisation systems

Metal oxide curing is used for non-olefin rubbers, like polychloroprene and chlorosulphonated polyethylene, which contain some active group (halogen or carboxyl
group) through which they can be cross-linked [1, 4]. In the case of polychloroprene,
the double bond is hindered by the neighbouring chlorine atom and hence the vulcanisation by sulphur as well. Therefore, neoprenes are generally vulcanised by reaction with zinc oxide, used along with magnesium oxide since the latter is necessary
to give the scorch resistance. Lead oxide and red lead are used as vulcanising agents
for polychloroprene when lower absorption and higher acid resistance are needed.

Cross-linking of butyl rubber with p-quinone dioxime proceeds through an oxidation step in the presence of red lead via the formation of dinitrosobenzene as an
intermediate, followed by the addition of this intermediate to two molecules of rubber with abstraction of two hydrogen radicals, which react with the polymer or with
more nitrosobenzene when quinone dioxime is regenerated.

Phenol formaldehyde resins with para-alkyl substitutes can be used as vulcanising agents for olefin rubbers such as EPDM, IIR and FKM. Resin cure gives very

thermal stable C–C cross-links. Vulcanisation with resin alone is a very slow process even at high temperatures, but the addition of catalysts such as $FeCl_3$, $SnCl_2$ or polymers, which liberate halogen acids, increases the curing rate.

References

[1]   Coran, AY. Science and technology of rubber. Academic Press, New York, USA, 1978.
[2]   Funke, W. Polymer yearbook. Hans-Georg Elias, R.A. Pethrick, Eds., Harwood Academic Publishers, New York, USA, 1984.
[3]   Hepburn, C, Reynolds, RJW. Elastomers: criteria for engineering design. Elsevier Applied Science Publishers, London, UK, 1979.
[4]   Fisher, HL. Chemistry of natural and synthetic rubbers. Reinhold Publishing Co., New York, USA, 1957.
[5]   Jensen WB. Lewis Acid-Base Interactions and Adhesion Theory. Rubber Chem. Technol. 1982, 55, 881.
[6]   Hofmann, W. Rubber technology handbook. Hanser Publishers, New York, USA, 1989.
[7]   Hofmann, W. Vulcanisation and vulcanising agents. McLaren, London, UK, 1967.
[8]   Das, PK, Datta, RN, Basu, PK. Cure modification effected by cycloalkylthioamines in the vulcanization of NR accelerated by thiocarbamyl sulfenamides and dibenzothiazyl disulfide. Rubber Chem. Technol. 1988, 61, 760.
[9]   Dogadkin, BA, Feldshein, MS, Belyaeva, EV. The action of binary accelerator systems of vulcanization. Rubber Chem. Technol. 1960, 33, 373.
[10]   Dogadkin, BA, Feldshein, MS, Belyaeva, EV. Chemical interaction and mechanism of activation of binary systems of vulcanization accelerators. J Polym Sci 1961, 53, 225.
[11]   Dogadkin, BA, Shershnev, VA. Vulcanization of rubber in the presence of organic accelerators. Rubber Chem. Technol. 1962, 35, 1.
[12]   Zaper, AM, Koenig, JL. Solid-state 13C NMR studies of vulcanized elastomers, 4. Sulfur-vulcanized polybutadiene. Makromol. Chem. 1988, 189, 1239.
[13]   Zaper, AM, Koenig, JL. Solid state carbon-13 NMR studies of vulcanized elastomers. II, Sulfur vulcanization of natural rubber. Rubber Chem. Technol. 1987, 60, 252.
[14]   Krejsa, MR, Koenig, JL. A review of sulfur crosslinking fundamentals for accelerated and unaccelerated vulcanization. Rubber Chem. Technol. 1993, 66, 376.
[15]   Wolfe JR, Pugh JL, Killian AS. The Chemistry of Sulfur Curing. III. Effects of Zinc Oxide on the Mechanism of the Reaction of Cyclohexene with Sulfur. Rubber Chem. Technol. 1968, 41, 1329.
[16]   Clough, RS, Koenig, JL. Solid-State Carbon-13 NMR Studies of Vulcanized Elastomers. VII. Sulfur-Vulcanized Cis-1,4 Polybutadiene at 75.5 Mhz. Rubber Chem. Technol. 1989, 62, 908.
[17]   Devlin, EF, Mengel, AL. Cis/trans isomerization in cured, black-reinforced, high cis-1,4-polybutadiene. J. Polym. Chem. 1984, 22, 843.
[18]   Coran, AY. Vulcanization: conventional and dynamic. Rubber Chem. Technol. 1995, 68, 351.
[19]   Palys, LH, Callais, PA. Understanding organic peroxides to obtain optimal crosslinking performance. Rubber World. 2003, 229, 35.
[20]   Garcia-Quesada, JC, Gilbert, M. Peroxide crosslinking of unplasticized poly (vinyl chloride). J. Appl. Polym. Sci. 2000, 77, 2657.
[21]   Busci, A, Szocs, F. Kinetics of radical generation in PVC with dibenzoyl peroxide utilizing high-pressure technique. Macromol. Chem. Phys. 2000, 201, 435.

4 Elastomers

4.1 Elastomer overview

Over the last century, a large number of basic and specialty elastomers have been developed to meet a wide field of applications and operating environments [1, 2]. Their properties vary broadly in terms of elasticity, temperature range, strength, hardness, compatibility, environmental resistance and costs. These basic properties can be substantially modified by compound design.

As described in Chapter 1, prior to World War II, developments were being actively pursued in Germany in the production of a synthetic polymer as a replacement for natural rubber (NR) in general-purpose applications. Through commercial contacts between German and American manufacturers, much detail of these materials and their manufacture was known in the USA as well. Hence, as a wartime necessity to make up for the deficiency of NR supplies to the allies, large-scale manufacture of the styrene–butadiene polymers with a 25% styrene and 75% butadiene content in the USA began. From this period, big efforts were made worldwide in developing new synthetic rubbers for both general purpose and special uses.

General-purpose elastomers, often referred to as commodity elastomers, are easily recognised in the market because of their big volume of usage and their lower selling prices with respect to specialty elastomers. General-purpose elastomers are hydrocarbon polymers, including styrene–butadiene rubber (SBR), butadiene rubber (BR) and polyisoprene rubber [both NR and IR (synthetic, isoprene rubber)]. These diene rubbers contain chemical unsaturation in their backbones, causing them to be rather susceptible to the attack by oxygen and especially by ozone. Additionally, they are readily swollen by hydrocarbon fluids. The primary application of these elastomers is in automotive and truck tyres, thanks to the high strength, good abrasion resistance, low hysteresis and high resilience.

Since in many applications, general-purpose elastomers are unsuitable due to their insufficient resistance to swelling, aging and/or elevated temperatures, specialty elastomers have been developed to meet these needs. Special-purpose rubbers are produced in much smaller quantities and have, besides a higher cost, a different degree of oil and solvent resistance and heat resistance with respect to those of the general-purpose class, which are produced in large quantities to replace NR with which they are comparable in non-oil-resistant properties.

Initially, special-purpose rubbers comprised only neoprene and hydrogenated acrylonitrile–butadiene rubbers (NBR and HNBR), which remain the workhorses because of their cost and their oil resistance. The market for neoprene rubbers has been much widened by the exploitation of their excellent resistance to ozone and weather, and by their use in fire-resistant application such as cable sheathing and conveyor belting for mines. The largest outlets for nitrile rubbers are in the engineering

https://doi.org/10.1515/9783110640328-004

industries for oil seals, o-rings, gaskets, fuel and oil hoses. Later, chlorosulphonated polyethylene rubbers have been developed and established for applications where solvent, chemical, ozone and weathering resistance are required as well. Fluorocarbon rubbers, with inferior low-temperature properties with respect to nitrile rubber, but having superior oil and heat resistance, represent a relevant improvement among the specialty rubbers. They are widely applied in the aircraft, aerospace and automotive industries. The high price of both fluorocarbon and silicone rubbers restricts their widespread use, even though silicones are unique in their varied range of service temperature, thus representing one of the most used special-purpose materials.

In Tab. 4.1, the most common elastomers are listed, together with the corresponding trade names and abbreviations. NR, polyisoprene, polybutadiene, SBR, NBR, EPDM (ethylene–propylene–diene monomer) and butyl rubber are usually vulcanised by sulphur. Silicone and fluoroelastomers are cured by peroxides, whereas neoprene is vulcanised with magnesium oxide. In Tab. 4.2, the most

Tab. 4.1: Trade names and abbreviations of the most common elastomers.

Elastomer	Trade name	Abbreviation	
		ISO 1629	ASTM 1418
Acrylonitrile–butadiene rubber (nitrile rubber)	Europrene Krynac Perbunan NT Breon	NBR	NBR
Hydrogenated acrylonitrile–butadiene rubber	Therban Zetpol	HNBR	HNBR
Polyacrylate rubber	Noxtite Hytemp Nipol AR	ACM	ACM
Chloroprene rubber	Baypren Neoprene	CR	CR
Ethylene–propylene–diene monomer rubber	Dutral Keltan Vistalon Buna EP	EPDM	EPDM
Silicone	Elastoseal Rhodorsil Silastic Silopren Addisil	VMQ	VMQ
Fluorosilicone	Silastic	FVMQ	FVMQ

Tab. 4.1 (continued)

Elastomer	Trade name	Abbreviation	
		ISO 1629	ASTM 1418
Tetrafluoroethylene–propylene copolymer elastomer	Aflas	FEPM	TFE
Butyl rubber	Butyl	IIR	IIR
Styrene–butadiene rubber	Buna S Europrene Polysar	SBR	SBR
Natural rubber	–	NR	WR
Fluorocarbon rubber	Fluorel Tecnoflon Viton	FKM	FKM
Perfluoro rubber	Kalrez Isolast	FFKM	FFKM
Polyurethane	Desmopan Elastollan Adiprene	TPU	TPU
Chlorosulphonated polyethylene rubber	Hypalon	CSM	CSM
Polysulphide elastomer	Thiokol	OT/EOT	TWT
Epichlorohydrin elastomer	Hydrin	CO/ECO	CO/ECO

important types of synthetic rubbers are collected on the basis of the group classification. It is evident that rubber compounders have a wide spectrum of elastomers to choose from, to meet one or more of the requirements for specific end-uses.

Rubbers can be broadly shared into two families: thermosets and thermoplastics. Thermosets are 3D molecular networks with the macromolecular chains held together by chemical bonds. They absorb solvent and swell, but do not dissolve; furthermore, they cannot be reprocessed simply by heating. The macromolecular chains of thermoplastic elastomers are not connected by primary chemical bonds, but they are joined by the physical aggregation of parts of the molecules into hard domains. Hence, thermoplastic rubbers dissolve in suitable solvents and soften on heating, so they can be processed repeatedly. In some cases, thermoplastic and thermoset rubbers might be used interchangeably, even if in demanding uses (i.e., tyres, engine mounts, etc.) thermoset elastomers are exclusively employed because of their better elasticity, resistance to set and durability.

Tab. 4.2: The grouping of the most important types of rubber.

Group	Character	Rubber
M	Saturated carbon molecules in the main macromolecular chain	Polyacrylate rubber Ethylene acrylate rubber Chlorosulphonated polyethylene rubber Ethylene propylene diene rubber Ethylene propylene rubber Fluorocarbon rubber Perfluoro rubber
O	With oxygen molecules in the main macromolecular chain	Epichlorohydrin rubber Epichlorohydrin copolymer rubber
R	Unsaturated hydrogen–carbon chain	Chloroprene rubber Butyl rubber Nitrile butadiene rubber Natural rubber Styrene–butadiene rubber Hydrogenated nitrile–butadiene rubber
Q	With silicone in the main chain	Silicone rubber Fluorosilicone rubber
U	With carbon, oxygen and nitrogen in the main chain	Polyester urethane Polyether urethane

Sections 4.1.1 and 4.1.2 deal briefly on both thermoset and thermoplastic rubbery materials available on the market, giving a quick overview of their peculiar characteristics. Some of the most used thermoset elastomers will be widely described later in this chapter.

4.1.1 Thermoset elastomers

Acrylic (ACM) (alkyl acrylate copolymer)

ACMs are copolymers of a dominant acrylate monomer (ethyl or butyl) and a cure-site monomer, such as 2-chloroethyl vinyl ether. Butyl acrylate results in a lower T_g, but poorer oil resistance compared to ethyl acrylate. Copolymerisation with acrylonitrile (ACN) improves the oil resistance. They show an outstanding resistance to oil and oxygen at normal and high temperatures, good weathering and ozone resistance, poor resistance to moisture, acids and bases. They are commonly used in automotive transmission seals and hoses and in adhesive formulations.

Bromobutyl (BIIR)

It is a butyl rubber modified by the introduction of a small amount of bromine, giving improved ozone and environmental resistance, stability at high temperatures and compatibility with other diene rubbers in blends. It shows also increased adhesion to other rubbers and metals. Its properties are similar to those of chlorobutyl rubber.

Butadiene (BR) (polybutadiene)

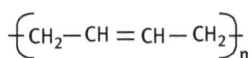

$$\left[CH_2{-}CH{=}CH{-}CH_2 \right]_n$$

BR (T_g = −50 °C) is widely used in blends with NR and SBR for tyres, where it reduces the heat build-up and improves the abrasion resistance. It shows low hysteresis, good flexibility at low temperatures and high abrasion resistance in severe conditions. It is used in shoes, conveyor and transmission belts.

Butyl (IIR) (isobutylene–isoprene copolymer)

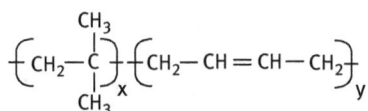

$$\left[CH_2{-}\overset{\displaystyle CH_3}{\underset{\displaystyle CH_3}{C}} \right]_x \left[CH_2{-}CH{=}CH{-}CH_2 \right]_y$$

Butyl rubber is a copolymer of isobutylene with a small percentage of isoprene to provide the sites for curing. IIR has unusually low resilience for an elastomer with such a low T_g (−65 °C). Because IIR is largely saturated, it has excellent aging stability. It shows high damping at ambient temperatures, good ozone, heat and chemical resistance, but not to oils. Thanks to its low permeability to gases, it is used for inner tubes and gas masks. Other uses include wire and cable applications, pharmaceutical closures and vibration isolation.

Chlorinated polyethylene (CPE)

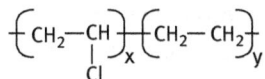

$$\left[CH_2{-}\underset{\displaystyle Cl}{CH} \right]_x \left[CH_2{-}CH_2 \right]_y$$

It is produced by chlorination of polyethylene having a chlorine content varying from 25 to 42 wt% (typically around 36 wt%). The saturated backbone and the increase of the chlorine content improve oil, fuel and flame resistance, but result in poorer heat resistance. Chlorinated polyethylene shows good chemical resistance to hydrocarbon fluids and to elevated temperatures, but poor mechanical strength. Indeed, the mechanical properties may deteriorate above 100 °C. It is used in the wire and cable industry, as well as for pond liners.

Chlorobutyl (CIIR)
It is a butyl rubber modified by the introduction of a small amount of chlorine, giving improved ozone and environmental resistance, stability at high temperatures and compatibility with other diene rubbers in blends. It shows increased adhesion to other rubbers and metals and similar properties to those of bromobutyl rubber.

Chlorosulphonated polyethylene (CSM)

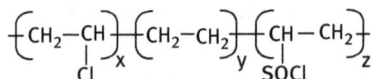

$$\left(CH_2-\underset{\underset{Cl}{|}}{CH}\right)_x \left(CH_2-CH_2\right)_y \left(\underset{\underset{SOCl}{|}}{CH}-CH_2\right)_z$$

Commercial grades contain from 25 to 45 wt% of chlorine and 1–1.5 wt% of sulphur. This elastomer show low gas permeability and good resistance to oxygen, ozone, light, weathering and flame resistance. The oil resistance increases with increasing the chlorine content, while the low-temperature flexibility and the heat aging resistance are improved when the chlorine content is low. Its excellent UV stability makes it useful as roof sheeting and for pond liners, as well as wire and cable applications, coated fabrics and hoses.

Epichlorohydrin (CO/ECO)

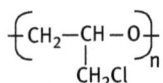

$$\left(CH_2-\underset{\underset{CH_2Cl}{|}}{CH}-O\right)_n$$

Polyepichlorohydrin (CO) and its copolymers with ethylene oxide (ECO) which have lower T_g (−40 °C) show high resistance to aging, oxidation, ozone and hot oil, good resistance to hydrocarbon solvents, moderate low temperature flexibility, poor electrical properties and abrasion resistance. They can be attacked by strong mineral and oxidising agents and chlorine. Their main use is in the automotive sector for seals, hoses, gaskets and o-rings. Thanks to the presence of chlorine, they show excellent resistance to fuel, oil and flame.

Ethylene propylene (EPM/EPDM)

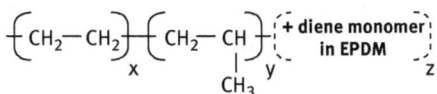

$$\left[CH_2-CH_2 \right]_x \left[\underset{\underset{CH_3}{|}}{CH_2-CH} \right]_y \left[\underset{\text{in EPDM}}{\text{+ diene monomer}} \right]_z$$

The ratio of ethylene to propylene in commercial grades varies from 50/50 to 75/25. Typical T_g is around −60 °C. Both EPM and EPDM have excellent ozone and weathering resistance, excellent hot water and steam resistance, good resistance to inorganic and polar organic chemicals, but low resistance to hydrocarbons.

Fluorocarbon (FKM)

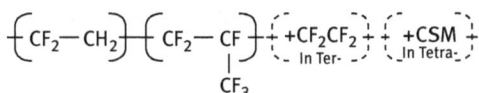

$$\left[CF_2-CH_2 \right] \left[\underset{\underset{CF_3}{|}}{CF_2-CF} \right] \left[\underset{\text{In Ter-}}{+CF_2CF_2} \right] \left[\underset{\text{In Tetra-}}{+CSM} \right]$$

Fluorocarbon rubbers are among the most inert and expensive elastomers (CSM = cure-site monomer). Their densities are around 2 g/cm^3 and the service temperature can exceed 250 °C. Properties vary significantly with type. T_g of polyvinylfluoride is −35 °C and that of polyhexafluoropropylene is +11 °C and for polytetrafluoroethylene is −130 °C. Generally, they show excellent ozone and weathering resistance and good heat resistance.

Fluorosilicone (FVMQ)

Thanks to their excellent resistance to oils, fuels and solvents, fluorine-containing silicones have found extensive applications in spite of their high price. They provide chemical properties similar to those of fluorinated organic elastomers, but their operational temperature range is wider than in the case of the former ones (−57 to 205 °C). Commercial fluorosilicones usually contain a small amount (about 0.2 wt%) of methyl vinyl siloxane as a cure-site monomer, while the fluorosilicone component may range from 40 to 90 wt%. Fluorosilicone compounds have never been more popular as they are today, especially for applications that require low- and high-temperature performances in contact with jet and automotive fuels, many solvents and engine oils. Nowadays, they have found an increased use in many high-volume automotive applications.

Hydrogenated nitrile (HNBR)

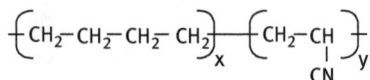

$$\left[CH_2-CH_2-CH_2-CH_2 \right]_x \left[\underset{\underset{CN}{|}}{CH_2-CH} \right]_y$$

Derived from conventional nitrile rubber by hydrogenation of the unsaturated bonds in the butadiene unit of the polymer, HNBR shows good oil/fuel and chemical resistance and good weathering resistance. Mechanical properties including tensile strength, tear, modulus, elongation at break and abrasion are excellent. It can be used in a wide temperature range (−40 to +160 °C) and can be compounded for excellent resistance to rapid gas decompression. Disadvantages include cost and limited resistance to aromatic solvents. As with nitrile, many properties can be influenced by varying the ACN/butadiene ratio.

Synthetic cis-polyisoprene (IR)

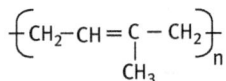

$$\left[CH_2-CH=\underset{\underset{CH_3}{|}}{C}-CH_2 \right]_n$$

The chemical structure is similar to that of NR, but it is less easy to process and it has lower tensile and tear strength. Its relative purity provides better performances at lower temperatures ($T_g = -70$ °C).

Natural rubber (NR)

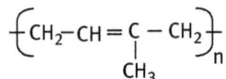

$$\left[CH_2-CH=\underset{\underset{CH_3}{|}}{C}-CH_2 \right]_n$$

It has high resilience and tensile strength, good abrasion resistance, low cost but poor oil and weathering resistance. The typical temperature range is from −50 to + 100 °C. Also T_g of NR is around −70 °C. NR can be used with some dilute inorganic chemicals and polar organics.

Nitrile (NBR)

$$\left[CH_2-CH=CH-CH_2 \right]_x \left[CH_2-\underset{\underset{CN}{|}}{CH} \right]_y$$

Butadiene   **Acrylonitrile**

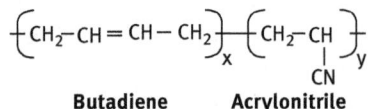

As with hydrogenated nitrile, many properties of NBR can be influenced by varying the ACN/butadiene ratio. NBR shows good aliphatic hydrocarbon oil and fuel resistance, high resilience, limited weathering resistance and only modest temperature resistance. The typical temperature range is from −30 to + 120 °C, even if low-temperature grades available down to −50 °C.

Perfluorocarbon (FFKM)

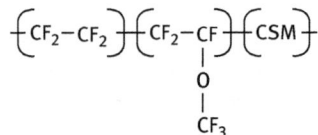

$$\left[CF_2-CF_2 \right]\left[CF_2-\underset{\underset{\underset{\underset{CF_3}{|}}{O}}{|}}{CF} \right]\left[CSM \right]$$

Some grades are suitable for continuous use at 327 °C, with chemical resistance being almost universal. However, their moderate mechanical properties deteriorate rapidly at elevated temperatures and at temperatures below 0 °C. FFKM is very expensive.

Polychloroprene or neoprene (CR)

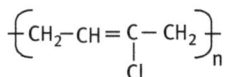

Due to the polar nature of molecules thanks to the presence of chlorine atom, neoprene shows very good resistance to oils and it is flame resistant. It has good weather and ozone resistance, low cost and limited temperature resistance. The typical temperature range is from −40 to +120 °C (T_g = −65 °C).

Polysulphide (OT/EOT)

Polysulphide rubbers are condensates of sodium polysulphides with organic dihalides. They contain a substantial proportion of sulphur in their structure (up to 80%). The adjacent ethylene and sulphide units give a very stiff chain. Flexibility can be increased by addition of ethylene oxide (T_g = −27 °C). Since their commercial introduction in 1929 by Thiokol Chemical Corporation, they have been utilised in specialty applications due to their excellent oil, fuels and solvent resistance as well as good aging properties. Although the original polymers were solid rubbery materials, today the predominant product, discovered some 20 years later, is the mercaptan-terminated liquid polymer. It can be transformed in situ from a liquid state into a solid elastomer, even at low temperatures, which makes its use convenient for adhesives, coatings and sealants where good resistance to solvents is required. Polysulphide rubbers have very good resistance to oxygen and ozone, no permeability to gases but they have poor mechanical properties and poor heat resistance. Because of their excellent oil and solvent resistance and impermeability to gases, polysulphides find applications in specialty areas, such as the manufacture of rollers for can lacquering, quick drying printing ink application and grain coating of paint on metals. They are blended with other synthetic rubbers for improved processing.

Silicone (VMQ)

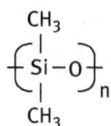

Silicone shows moderate physical properties but it is able to retain them over a very wide temperature range (T_g = −130 °C). Therefore, it is widely used in many sectors such as pharmaceutical, medical, wire and cable, automotive and aerospace.

Styrene–butadiene rubber (SBR)

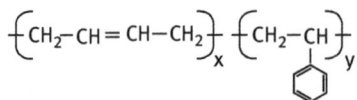

$$\left[CH_2-CH=CH-CH_2 \right]_x \left[CH_2-CH \right]_y$$

SBR is widely used in car and light vehicle tyres and also for conveyor belts, moulded rubber goods, shoe soles and roll coverings. T_g for a typical 75/25 blend is around −60 °C.

Tetrafluoroethylene–propylene (FEPM)

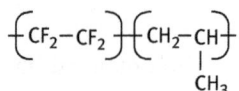

$$\left[CF_2-CF_2 \right]\left[CH_2-CH \right]$$
$$CH_3$$

The original developer (Dupont in 1960s) never commercialised the elastomer, a copolymer of propylene and tetrafluoroethylene (TFE), due to the lack of market interest. Reintroduced by Asahi Glass Co. in 1970s, the elastomer found a specific niche in the emerging high-temperature 'sour' oil and gas production environment, operating continuously at temperatures in excess of 200 °C in the presence of hydrocarbons, carbon dioxide, hydrogen sulphide and brines. The particular properties of propylene–TFE rubber are due to the presence in its structure of a non-substituted alkene and a totally fluorinated alkene. This category of elastomers is unique in both aqueous and no aqueous electrolytes, which include engine coolants and highly stabilised lubricants. Along with the FFKM category, FEPM is suitable for service in all classes of petroleum-based hydraulic fluids, polyesters, phosphate ester and silicic acid ester fluids.

4.1.2 Thermoplastic elastomers

Thermoplastic elastomers have many of the physical properties of vulcanised rubbers but they can be processed as thermoplastics. Since their commercial introduction in the 1960s, they have become a significant part of the elastomer industry and used in many applications as adhesives, footwear, medical devices, automobile parts and asphalt modification. Their disadvantages are the relatively high cost of raw materials, the poor chemical and heat resistance, the high compression set and the low thermal stability. For more details on thermoplastic elastomers, see Chapter 5.

Styrenic block copolymers (SBS, SIS, SEBS)
Styrenic block copolymers are the largest volume and lowest priced member of
the thermoplastic elastomer family. They are readily mixed with other polymers,
oils and fillers, enabling the versatile tuning of product properties. They are used
in enhancing the performance of bitumen in road paving and in roofing applications, particularly under extreme weather conditions. They are also widely
applied in adhesives, sealants, coatings and footwear.

Polyurethane

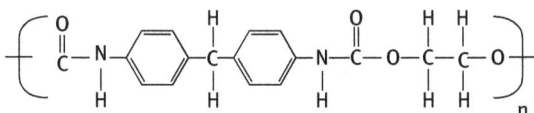

Thermoplastic polyurethanes are a very versatile type of materials, available in a more
limited hardness range than styrenics, and characterised by excellent strength and
toughness, good abrasion resistance, high tensile and tear strength, good resistance to
aliphatic solvents and mineral oils, oxygen and ozone. Polyurethane shows poor heat
and creep resistance, particularly in moist conditions. Of the two major types, polyester and polyether, the latter has better hydrolytic stability and low-temperature performances. The electrical properties of the polyurethanes are not good enough for use
as primary insulation, but their general toughness leads to their use in cable jacketing.
Other uses include fabric coatings, bellows and automotive body components.

Copolyether ester elastomers
These materials are strong, tough and oil resistant, but are only available in a limited hardness range. They are also resistant to oxygen and ozone. They are used in
moulded goods applications requiring exceptional toughness and flex resistance
together with moderate heat and chemical resistance. Applications include cable
jackets, tubing, automotive bellows, gear wheels and business machine parts.

Polyester amide elastomers
They have similar properties to copolyether ester elastomers, except service temperatures that are lower. They show good strength and toughness as well as oil, oxygen and ozone resistance. The hardness range is limited as the hydrolytic stability.

4.2 Silicone

Because of its unique properties and somewhat higher price compared to the other
common elastomers, silicone rubber is usually classed as a specialty elastomer,

although it is increasingly used as a cost-effective alternative in a variety of applications [3–5].

One of the most basic technical errors made by people referring to this material is confusing silicon with silicone. The former silicon is used to denote the elemental material, Si. The latter refers to materials in which silicon is bonded to oxygen. Silicon is the most elemental raw material from which all silicone chemistry finds its roots.

The term silicone was coined by F. S. Kipping in 1901 and refers to the formal analogy between these silicon compounds and the equivalent oxygen compounds of carbon (polyketones). By analogy with ketones, the name silicone describes compounds having the brut formula R_2SiO.

The basic silicone polymer is polydimethylsiloxane with a backbone of silicon–oxygen links and two methyl groups on each silicon. The Si–O–Si group is better described by the term *siloxane*.

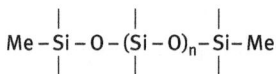

$$Me-\underset{|}{\overset{|}{Si}}-O-(\underset{|}{\overset{|}{Si}}-O)_n-\underset{|}{\overset{|}{Si}}-Me$$

The silicon–oxygen backbone provides a high degree of inertness to ozone, oxygen, heat (up to 315 °C), UV light, moisture and general weathering effects, while the methyl substituents confer a high degree of flexibility. Again, the simultaneous presence of organic groups attached to an inorganic backbone gives to silicones a combination of unique properties and allows their use in fields as different as aerospace (low- and high-temperature performance), electronics (electrical insulation), health care (excellent biocompatibility) or building industries (resistance to weathering).

The chemical structure of silicones allows them to be produced in a number of variations. By using siloxane units with different valences, products can be made with oily, polymeric, resinous or cross-linked properties. At the same time, the organic groups bound to the silicon pave the way for a diverse range of modifications. Indeed, the methyl groups along the chain can be substituted by many others (e.g., phenyl, vinyl or trifluoropropyl) so modifying the basic properties of silicone. It is this variability that makes possible the impressive variety of silicone products, such as greases, release agents, antifoam agents, paint additives, paper coatings, hydrophobising agents and high- or room-temperature vulcanising silicone rubbers.

According to ASTM D1418 there are various classes of silicone rubbers [6]:

– MQ: silicone rubbers having only methyl groups on the polymer chain (polydimethylsiloxanes), especially for coating purposes
– VMQ: silicone rubbers of general purpose having methyl and vinyl groups on the polymer backbone
– PMQ: silicone rubbers having methyl and phenyl substitutions on the polymer chain, useful for low-temperature applications

- PVMQ: silicone rubbers having methyl, phenyl and vinyl groups on the polymer backbone, used in extremely low-temperature applications
- FVMQ: silicone rubbers having fluorine, methyl and vinyl substitutions on the polymer chain, used in applications involving fuel, oil and solvent resistance

Of the available silicone elastomers, the methyl and vinyl types are the most widely used. Vinyl groups increase the reactivity of the polymer and provide much faster vulcanisation and more elastic vulcanisates.

There are three main industrial classifications of silicone rubbers:
- *High-temperature vulcanising silicone (HTV):* also called heat-curable or hard siliocone. These are usually in a semi-solid rubber-like form in the uncured state. Solid silicone rubbers are cured at elevated temperature, either by means of organic peroxides or platinum catalysts. They require rubber-type processing to produce finished items.
- *Room-temperature vulcanising silicone (RTV):* they usually come as a flowable liquid and are used for sealants, mould making, encapsulation and potting. RTV silicones are low-molecular-weight polydimethylsiloxane liquids with reactive end groups. As with the heat-cured polymers, they can contain phenyls for improved low-temperature flexibility or fluoroalkyl groups, for improved oil and solvent resistance and even for broader temperature service. Vulcanisation of the RTV silicones occurs by either condensation or addition reaction. Condensation cures can be independent or dependent on moisture. These silicones are not generally used as conventional rubbers.
- *Liquid silicone rubbers:* also called heat-curable liquid materials. They are characterised by a lower viscosity compared to the solid silicone rubbers. Their consistency and curing mechanism give them outstanding processing advantages. Liquid silicone rubbers are two-component compounds (A + B) supplied as ready to be processed: cure starts after mixing the two separate parts. Component A contains a Pt catalyst and component B a SiH-functional curing agent. They are vulcanised by addition curing. Physical properties of liquid silicone rubbers are comparable to general-purpose grades and high-strength peroxide-cured elastomers. Compared to peroxide curing, liquid silicone rubbers do not release any curing by-products. Post-curing is therefore not usually required except for reaching the optimal dimensional stability and the required physical–mechanical characteristics in vulcanisates. In comparison with hard silicone rubber, cycle times are shorter, thanks to the faster curing rate and to the low viscosity. Indeed, curing is often completed in a few seconds at temperatures of about 200 °C, depending on the shape and size of a given item. Moulds can have more than 250 cavities and complex part geometry can be obtained requiring no secondary finishing, becoming an advantage with respect to the use of HTV for particular applications.

4.2.1 From sand to silicones

The chemistry of silicon and its compounds is extremely dynamic [3, 7]. Hardly any technology in recent decades has shaped technical progresses so strikingly as silicon chemistry. This success story begins with the element silicon. In 1940–41, Professors Müller and Rochow independently discovered how to react silicon with methyl chloride gas (CH_3Cl) to form liquid methylchlorosilanes. This step provided the starting materials, which are silanes, for the industrial manufacturing of silicones and launched a global boom in silicone production.

Silicone polymers are obtained by a three-step process:

1. Chlorosilane synthesis

Silicon metal is obtained from the reduction of sand at high temperature:

$$SiO_2 + 2C \rightarrow Si + 2CO$$

Methyl chloride is produced by condensation of methanol with hydrochloric acid:

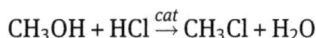

$$CH_3OH + HCl \overset{cat}{\rightarrow} CH_3Cl + H_2O$$

The solid–gas reaction giving chlorosilanes is quite complex. It takes place in a fluidised bed of silicon metal powder in which flows a stream of methyl chloride, usually at temperatures from 250 to 350 °C and pressures from 1 to 5 bars. A copper-based catalyst is used. In order to maximise the reaction efficiency, the solid silicon must contain low concentrations of other metallic components such as Cu, Fe, Al and Ca. The reaction mass needs to be heated, but once the reaction temperature is reached, the reaction becomes exothermic, and consequently requires a very stringent temperature control. The exothermic reaction has a yield of 85–90%.

The reaction mechanism is not completely understood yet: chemisorption phenomena on active sites seem to be preferred to the radical-based mechanism originally proposed. At the end, a mixture of different silanes is produced:

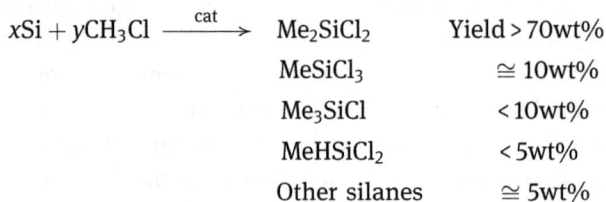

$$xSi + yCH_3Cl \xrightarrow{\text{cat}}$$

Me_2SiCl_2	Yield > 70wt%
$MeSiCl_3$	$\cong 10$wt%
Me_3SiCl	< 10wt%
$MeHSiCl_2$	< 5wt%
Other silanes	$\cong 5$wt%

The various silanes are separated by distillation: as the boiling points are very close, long distillation columns are always seen in the silicone factories. Silanes are colourless, clear, mobile liquids, and are soluble in organic solvents. Their low molecular mass makes them highly volatile.

Dimethyldichlorosilane is separated since it is the basic monomer for the synthesis of polydimethylsiloxanes. Redistribution reactions can be used to convert the other silanes, thus increasing the commercial yield of the production equipment. Because of the complicated process technology and the high capital requirements to construct plants suitable to practice the chemistry, few companies worldwide actually carry out the Rochow process.

2. Chlorosilane hydrolysis

Polydimethylsiloxanes are obtained by the hydrolysis of dimethydichlorosilane in the presence of excess of water according to the following reaction:

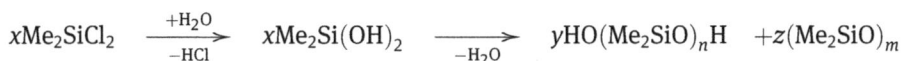

$$x\text{Me}_2\text{SiCl}_2 \xrightarrow[-\text{HCl}]{+\text{H}_2\text{O}} x\text{Me}_2\text{Si(OH)}_2 \xrightarrow[-\text{H}_2\text{O}]{} y\text{HO(Me}_2\text{SiO)}_n\text{H} + z(\text{Me}_2\text{SiO})_m$$

This heterogeneous and exothermic reaction gives formally a disilanol, $\text{Me}_2\text{Si(OH)}_2$, which readily condenses with HCl, acting as a catalyst, to give a mixture of linear $(\text{HO(Me}_2\text{SiO)}_n\text{H})$ or cyclic oligomers $((\text{Me}_2\text{SiO})_m)$ by inter- or intramolecular condensation. The ratio between the two kinds of oligomers depends on the hydrolysis conditions, such as concentrations, pH and solvents. The oligomeric mixture separates from the aqueous acid phase, so oligomers are water-washed, neutralised and dried. HCl is recycled and reacts with methanol to give the methyl chloride used in the first step of the process.

3. Polymerisation and polycondensation

The linear and cyclic oligomers obtained by hydrolysis of dimethyldichlorosilane have too short chains for most applications. Thus, they must be condensed (linears) or polymerised (cyclics) to give macromolecules of sufficient length for ensuring useful properties to the polymer [4].

Cyclics can be opened and polymerised to form long linear chains. The reaction is catalysed by many acid or base compounds, giving at the equilibrium a mixture of cyclic oligomers plus a distribution of polymers [6]. The proportion of cyclics depends on the substituents along the chain, the temperature and the presence of a solvent.

The catalyst removal (or neutralisation) is an important step in silicone preparation, since most of them may also catalyse depolymerisation, particularly in the presence of water traces at elevated temperatures. To benefit as much as possible from the thermal stability of the silicone, it is therefore essential to remove all remaining traces of the catalyst. Labile catalysts, which decompose or are volatilised above the optimum polymerisation temperature, have been developed. They are removed by brief overheating, thus avoiding the catalyst neutralisation or filtration.

4.2.2 Silicone elastomers

Silicone polymers are easily transformed into a three-dimensional elastomeric network thanks to a cross-linking reaction, which allows the formation of chemical bonds between adjacent macromolecular chains [4, 5]. This is achieved according to one of the following reactions.

1. Cross-linking with radicals

The efficient cross-linking with radicals is achieved only when some vinyl groups are present on the polymer chains. The following mechanism has been proposed for the cross-linking made by radicals generated from an organic peroxide [5]:

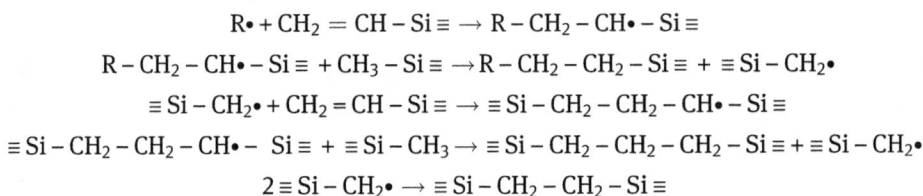

$$R\bullet + CH_2 = CH - Si \equiv \rightarrow R - CH_2 - CH\bullet - Si \equiv$$
$$R - CH_2 - CH\bullet - Si \equiv + CH_3 - Si \equiv \rightarrow R - CH_2 - CH_2 - Si \equiv + \equiv Si - CH_2\bullet$$
$$\equiv Si - CH_2\bullet + CH_2 = CH - Si \equiv \rightarrow \equiv Si - CH_2 - CH_2 - CH\bullet - Si \equiv$$
$$\equiv Si - CH_2 - CH_2 - CH\bullet - Si \equiv + \equiv Si - CH_3 \rightarrow \equiv Si - CH_2 - CH_2 - CH_2 - Si \equiv + \equiv Si - CH_2\bullet$$
$$2 \equiv Si - CH_2\bullet \rightarrow \equiv Si - CH_2 - CH_2 - Si \equiv$$

where $\equiv$ represents two methyl groups and the rest of the polymer chain.

This reaction is used for high consistency silicone rubbers such as those used in extrusion or injection moulding that are cross-linked at elevated temperatures. The peroxide is added before the compound processing. Cross-linking by organic peroxide takes place in the temperature range 110°–160 °C and it is generally quick. During cure, some precautions are needed to avoid the formation of voids by the volatile residues of the peroxide. Post-curing at 200°–250 °C for approximately 24 h is necessary to remove any peroxide residual and to develop the optimum properties, particularly the heat resistance and the dimensional stability. Post-curing of vulcanisates allows to remove also some volatiles, which might act as depolymerisation catalysts at high temperatures.

2. Cross-linking by condensation

This process is used in sealants such as the ones available from the 'do-it-yourself' shops, ready to use and requiring no mixing. These sealants are called one-part sealant (RTV-1) and require moisture as the second component. Cross-linking starts when the product is squeezed from the cartridge and comes into contact with moisture. Therefore, cross-linking requires that moisture diffuses within the product and the cure proceeds from the outside surface towards the inside. Atmospheric moisture is sufficient to trigger the reaction. Thickness should be limited if only one side is exposed to the moisture source. Curing is relatively slow depending on the moisture access into the polymer.

RTV-1 is formulated from a reactive polymer prepared from a hydroxy-end-blocked polydimethylsiloxane and a large excess of methyltriacetoxysilane:

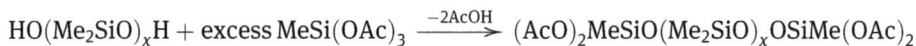

$$HO(Me_2SiO)_xH + excess\ MeSi(OAc)_3 \xrightarrow{-2AcOH} (AcO)_2MeSiO(Me_2SiO)_xOSiMe(OAc)_2$$

As a large excess of silane is used, the probability of two different chains reacting with the same silane molecule is remote and all the chains are end-blocked with two −OAc functions. The resulting product is still liquid and can be stored in sealed cartridges. Curing takes place when the cartridge is opened and the silicone comes into contact with the moisture of the air. Chemically, the acetoxy groups are hydrolysed to give silanols, which allow further condensation:

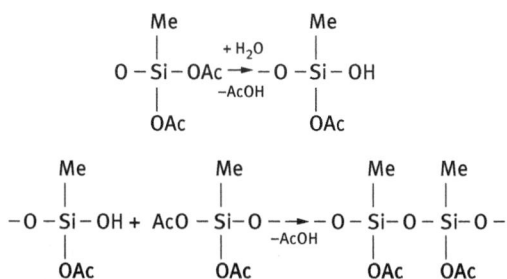

In this way, two chains have been linked, and the reaction proceeds further from the remaining acetoxy groups. An organometallic tin catalyst is normally used.

Acetic acid is released as a by-product of the process; thus, corrosion problems are possible on substrates such as concrete, with the formation of a water-soluble salt at the interface (and loss of adhesion at the first rain!). To overcome this, other systems have been developed. These include one-part sealants releasing less corrosive or non-corrosive by-products (e.g., oxime using the oximosilane $RSi(ON = CR'_2)_3$ or alcohol using the alkoxysilane $RSi(OR')_3$ instead of the above acetoxysilane).

Condensation curing in two-part systems (RTV-2) starts upon mixing the two components. Here, no atmospheric moisture is needed. Usually an organotin salt is used as catalyst but it limits the stability of the resulting elastomer at high temperatures. Alcohol is released as a by-product of the cure, leading to slight shrinkage upon cure. This precludes the fabrication of very precise objects due to the linear shrinkage (from 0.5% to 1%).

3. Cross-linking by addition
The above shrinkage problem can be overlapped, thanks to the addition process. Here, cross-linking occurs by the reaction between the vinyl-end-blocked polymers and the SiH groups carried by a functional oligomer:

$$\sim\!\sim\!\sim\!OMe_2Si - CH = CH_2 + H - Si \equiv \quad \xrightarrow{cat} \quad \sim\!\sim\!\sim\!OMe_2Si - CH_2 - CH_2 - Si \equiv$$

where $\equiv$ represents the remaining valences of Si.

Addition occurs mainly on the terminal carbon and is catalysed by Pt–metal complexes, preferably organometallic compounds, to enhance compatibility. Vulcanisation by Pt catalyst is very common for HTV silicone rubber. There is no by-product with this reaction; thus, the items produced with this cure mechanism are very accurate (no shrinkage) and vulcanise with good dimensional stability. However, the handling of this kind of silicone requires some precautions. Indeed, Pt in the complex may be easily bonded to electron-donating substances, like amine or organo-sulphur compounds, to form stable complexes with these poisons, making the catalyst inactive (inhibition). Consequently, cross-linking is hampered.

4.2.3 Physical–chemical properties of silicone

Silicones display the unusual combination of an inorganic chain, similar to silicates and often associated with high surface energy, with side organic groups that are often associated with low surface energy [4].

Siloxane chains exhibit high flexibility: rotation barriers are low and the siloxane chains can adopt many conformations. Rotation energy around a CH_2–CH_2 bond in polyethylene is 13.8 kJ/mol, but only 3.3 kJ/mol around a Me_2Si–O bond, corresponding to a nearly free rotation. The siloxane chains adopt a conformation that can be idealised by saying that the chain exposes a maximum number of methyl groups to the outside of the coil structure, while in hydrocarbon polymers, the relative backbone rigidity does not allow a selective exposure of the most organic or hydrophobic methyl groups. Therefore, the non-polar methyl groups located on the outside can rotate freely around the silicon–oxygen chain, forming a shield for the polar main chain. This shielding explains the low surface tension and gives to silicones their distinctive interfacial properties, including water repellence and good releasability.

The Si–O bonds are strongly polarised and without protection should lead to strong intermolecular interactions [5]. However, the organic groups only interact weakly with each other, shielded by the main chain. Thus, the chain–chain intermolecular force is weak and the coil formation capacity is high, resulting in high compressibility and excellent resistance to cold temperature.

The weak intermolecular interactions in silicones have other consequences. Glass transition temperatures are very low (i.e., around -130 °C for a polydimethylsiloxane compared to -70 °C for polyisobutylene, the analogue hydrocarbon). Thanks to their low surface tension (around 20 mN/m), silicones are able to wet most surfaces and themselves, a property that promotes good film formation and good surface covering in coating processes. The presence of a bigger free volume compared to hydrocarbons explains the high solubility and the high diffusion coefficient of gas into silicones. Therefore, silicones have a high permeability to oxygen, nitrogen and water vapour, even if liquid water is not able to wet a silicone surface.

The Si–O bond energy is significantly greater than that of a C–C bond (433 kJ/mol vs 355 kJ/mol). This has relevant effects on the stability and resistance of silicones. They show a remarkable thermal and thermo-oxidative resistance and provide electrical insulation. Properties such as volume resistivity, dielectric strength and power factor are not affected from changes in temperature; therefore, many types of wires and cables are insulated with silicone rubbers, since their excellent electrical properties are maintained at elevated temperatures. Silicone can be compounded to be electrically resistant, conductive or flame retardant, as well.

The thermal stability of silicones comes from the thermal stability of Si–O and Si–CH$_3$ bonds, which can be easily destroyed by concentrated acids and alkalis at ambient temperatures due to their partially ionic nature. When other groups in respect to methyl are present along the chain, the thermal stability is reduced, allowing to modify other properties. For example, a small percentage of phenyl groups along the chain perturbs sufficiently to affect crystallisation and allows the polymer to remain flexible at very low temperatures. Without sacrificing high-temperature properties, vinyl groups improve compression set resistance and facilitate vulcanisation.

4.2.4 Performances and applications

HTV silicone rubbers have a successful track record in practically all industries and new applications are added every day, thanks to their outstanding properties [8]. They are used in the automotive industry, in electrical applications, in food and personal hygiene, for machinery and in the construction industry, where often high temperature resistance and high thermal stability are required [5, 7]. These applications include a wide range of service temperatures (–50 to +250 °C, or even –90 to +300 °C for special formulations) in which the material properties remain constant, especially when excellent aging and weathering resistance are required. Indeed, silicone is unaffected by atmospheric exposure (sunlight, oxygen and moisture) and do not show ozone cracking.

In general, silicones are flexible at low temperatures due to their low glass transition temperature. The embrittlement point of conventional rubbers is between –20 and –30 °C compared to –70 °C of silicone, which in these environments remains elastic. Conversely, they can be used indefinitely at 150 °C with almost no change on their properties, so they are suitable for components that must suffer high-temperature environments.

At 205 °C, the silicone rubber has an estimated useful life from 2 to 5 years, depending on the end-use, while most organic materials will fail within a few days.

Silicone rubber performs unusually well when used in sealing applications since shows high resistance to hydrocarbons, oils and solvents. Over the entire temperature range of −85 to 360 °C, no available elastomer can match its low compression set.

Silicone has also antiadhesive and hydrophobic properties and low chemical reactivity as well. On the other hand, it generally offers poorer tensile, tear and abrasion properties than the most common organic rubbers, but this is routinely improved by reinforcement with fumed silica, which also enhances the electrical insulation properties.

Silicone rubbers are chemically inert, have no taste or smell and are, with few exceptions, physiologically acceptable to animal tissue. The very low toxicity of silicones explains their numerous applications, where the prolonged contact with the human body is involved, such as in textile fabrics, cosmetics, food and medical applications. These include baby bottle nipples, belts and hoses for conveying foods and food ingredients, medical tubings, surgical implants in human body and prosthetic devices. The excellent biocompatibility is partly due to the low chemical reactivity, low surface energy and hydrophobicity.

At 25 °C, the permeability of silicone rubber to gases is approximately 400 times that of butyl rubber. This allows this material to be used for applications such as the oxygen permeable membranes in medical devices.

4.3 Acrylonitrile/butadiene rubber

Nitrile rubber is commonly considered as the workhorse in the industrial and automotive industries [8–10]. With a temperature range from −40 to + 125 °C, NBR can withstand the most severe automotive applications requiring oil, fuel and chemical resistance, such as handling hose, seals and grommets. On the industrial side, NBR finds uses in roll covers, hydraulic hoses, conveyor belting, graphic arts, oil field packers and seals for all kinds of plumbing and water handling.

Under the abbreviation, NBR is comprised of a complex family of unsaturated polar copolymers of ACN and butadiene. The ACN content is the primary criterion for defining each specific NBR grade. It ranges from 20 to 50 wt% with 34% being a common and typical value. The ACN level, by reason of its polarity, determines several basic properties of nitrile rubbers. As the ACN content increases, oil resistance, solvent resistance, tensile strength, hardness, abrasion resistance, heat resistance, glass transition temperature, processability, cure rate with sulphur system and gas impermeability improve, but compression set resistance, resilience and low-temperature flex deteriorate.

As a general rule, raising the ACN level increases the compatibility with polar plastics such as PVC and also enables easier processing. Thanks to the PVC blending, nitrile rubbers can meet a broader range of physical or chemical requirements, such as improved abrasion, tensile, tear and flex properties and increased resistance to ozone and weathering. In non-black compounds the addition of the PVC resins also provides a greater pigment-carrying capacity that allows a better retention of pastel and bright colours.

By selecting an elastomer with the appropriate ACN content in balance with other properties, the rubber compounder can use NBR in a wide variety of application areas. The selection of the needed grade of NBR is generally based on oil resistance versus low-temperature performance. Polymers with high ACN content are used where the utmost oil resistance is required such as oil well parts, fuel cell liners, fuel hose and other applications requiring resistance to aromatic fuels, oils and solvents. Medium grades are used in applications where the oil is of lower aromatic content such as in petrol hose and seals. The low and medium-low ACN grades are used in cases where low-temperature flexibility is of greater importance than oil resistance.

Also the amount of butadiene shows a relevant effect on the end properties and performances of NBR. Commercial nitrile rubbers are available with ACN/butadiene ratios ranging from 18:82 to 45:55. This ratio is varied for specific oil and fuel resistance requirements. Blends of different grades are common to achieve the desired balance of properties.

In general, NBR is compounded along lines similar to those practiced with NR and SBR. NBR may be vulcanised by the conventional accelerated sulphur systems and also by peroxides. A tetramethyl thiuram monosulphide/sulphur cure is an excellent general-purpose system. Another widely used general-purpose cure system is 1.5 MBTS/1.5 sulphur.

Nitrile elastomers do not crystallise when stretched and so require reinforcing fillers to develop the optimum tensile strength, abrasion resistance and tear resistance. They also possess poor building tack and can suffer ozone cracking. Nitrile rubber has limited weathering resistance and poor aromatic oil resistance. Although nitrile rubbers are broadly oil and solvent resistant, they are susceptible to attack by certain strongly polar liquids, to which the non-polar rubbers, such as SBR or NR, are resistant. Nitrile rubber is poorly compatible with NR, but can be blended in all proportions with SBR. This decreases the overall oil resistance, but increases that to polar liquids in proportion to the SBR content.

4.3.1 Chemistry and manufacturing process

In 1930, in the course of their studies on the copolymerisation of 1,3-butadiene with monoolefins, Konrad and coworkers obtained a synthetic rubber based on butadiene and ACN which when vulcanised had excellent resistance to oil and petrol classifying it as a special-purpose rubber. Pilot plant production of Buna N, as this product was first named, started in Germany in 1934 and the full-scale production started in 1937 by Farbenfabriken Bayer AG (Germany) with the trade name PERBUNAN.

Currently, NBR is produced by emulsion polymerisation. Water, emulsifier, monomers (butadiene and ACN), radical-generating activator and other ingredients are introduced into the polymerisation vessels. The emulsion process yields a polymer latex that is coagulated using various materials (e.g., calcium chloride and aluminium sulphate) to form the crumb rubber that is dried and compressed into bales, ready for the next step of compounding.

As the ratio of butadiene to ACN in the polymer largely controls its properties, the design of the polymerisation recipe and the temperature at which this is carried out are important features of the nitrile rubber production. The nature and the amount of modifiers also influence the properties of the end polymer.

NBR producers vary the polymerisation temperature to make *hot* and *cold* polymers. The early nitrile rubbers were all polymerised at about 25–50 °C. These *hot* polymers contain a degree of branching in the polymer chains known as 'gel'. Since the early 1950s, by analogy with the developments in the emulsion polymerisation of SBR, an increasing number of nitrile rubbers started to be produced by *cold* polymerisation at about 5 °C. This resulted in more linear polymers containing little or no gel which were easier to process than *hot* polymers.

The current generation of *cold* NBRs spans a wide variety of compositions. ACN content ranges from 15% to 51%. *Cold* polymers are polymerised at a temperature range from 5 to 15 °C, depending on the balance of linear-to-branched configuration needed. Currently, *hot* NBR polymers are polymerised in the temperature range from 30 to 40 °C. The resulting highly branched polymers support good tack and strong bond in adhesive applications. The physically entangled structure of this kind of polymer also provides a significant improvement in hot tear strength compared with a *cold* polymerised counterpart. The natural resistance to flow makes *hot* polymers as excellent candidates for compression moulding and cellular rubber (i.e., sponge). Other applications are thin-walled or complex extrusions where the shape retention is important.

Third monomers (e.g., divinyl benzene and methacrylic acid) are added to the polymer backbone to provide advanced performances. *Cross-linked hot* NBRs are branched polymers that are further cross-linked by the addition of the difunctional monomer. These products are typically used in moulded parts to offer sufficient moulding forces, or back pressure, to eliminate the trapped air. Another use is to provide increased dimensional stability or shape retention for both extruded and calendered items. This leads to more efficient extrusion and vulcanisation of intricate shaped parts as well as improved release from calender rolls.

Addition of carboxylic acid groups to the NBR polymer backbone significantly alters both processing and curing properties. The result is a polymer matrix with significantly increased strength, measured by improved tensile, tear, modulus and abrasion resistance. The negative effects include the reduction in compression set, water resistance, resilience and some low-temperature properties.

Nitrile rubbers are available also with an antioxidant polymerised into the main polymer chain. The purpose is to provide additional protection to NBR during prolonged fluid service or in cyclic fluid and air exposure. Also the abrasion resistance is improved when compared with that of the conventional NBR, especially at elevated temperatures.

4.4 Hydrogenated nitrile butadiene rubber

Hydrogenated nitrile rubber derives from standard NBR in which the hydrogenation of the double residual bond of butadiene occurs [8–10]. The addition of hydrogen gas, in conjunction with a metal catalyst at well-defined temperature and pressure, brings about a selective hydrogenation to produce the highly saturated nitrile polymer. As the hydrogenation level increases, both the heat and ozone resistance improve.

The ACN content in HNBR ranges from approximately from 17 to 50 wt%. The ACN content not only controls the fluid and chemical resistance but also impacts the low-temperature performance. If the ACN content of the polymer increases, the volume swell of the associated compound decreases and the low-temperature flexibility becomes poorer. Therefore, for low-temperature performances, low ACN grades should be used. High-temperature performance can be obtained by using highly saturated HNBR grades with white fillers.

HNBR elastomers are typically cured with either peroxide or sulphur/sulphur donor-cure systems. Peroxide curing provides better compression set and heat resistance. When compared to SBR or NBR, the curing rate of HNBR tends to be slower. Therefore, to increase the curing rate a secondary accelerator should be employed in combination with the primary one.

Compounding techniques allow HNBR to be used over a broad temperature range (−40 to 165 °C), with minimal degradation over long periods of time. Depending on filler selection and loading, HNBR compounds typically have tensile strengths of 20–30 MPa when measured at 23 °C.

Hydrogenated nitrile elastomer compounds allow the manufacture of items offering peculiar and interesting properties such as:
- excellent mechanical characteristics
- high tensile strength and tear resistance
- excellent abrasion resistance
- good heat resistance (up to 150 °C) and elasticity at low temperatures
- good ozone and atmospheric agents resistance
- good compression set
- excellent lubricating oil resistance mainly with amine additives
- good resistance to liquids and lubricants for cooling circuits

Thanks to its excellent cost-performance balance for the most demanding applications, HNBR is the ideal choice in many cases. On a volume basis, the automotive market is the largest consumer, using HNBR for dynamic and static seals, hoses and belts. HNBR fills the gap left by NBR and FKM elastomers when high-temperature conditions require high tensile strength while maintaining excellent resistance to motor oil, sour gas, amine/oil mixtures, oxidised fuels and lubricating oils.

4.5 Fluoroelastomers (FKM)

Fluoroelastomers are a class of highly fluorinated polymers (fluorine content from 66 to 70 wt%) that provide extraordinary levels of resistance to chemicals, solvents, lubricants, hydraulic fluids, mineral acids, fuels, oils, heat, harsh chemicals and ozone with useful service life above 200 °C and a thermal stability up to 260 °C, superior to most all rubbers [11, 12]. The substitution of hydrogen with fluorine in the polymer backbone creates a unique class of rubbers, since the fluorine atoms provide the relative inertness of FKM elastomer.

Fluoro-olefin history began in 1892 with the work of Swarts, a Belgian chemist, who discovered the Swarts reaction, so named after him. This reaction led to the development of fluoroalkanes, the precursors of fluoroalkenes, which in turn led to halogenated refrigerants. Volume production of these refrigerant gases offered a readily available volume source of monomers necessary for the production of fluoroelastomers, which was discovered in the early 1950s.

The original fluoroelastomer was a copolymer of hexafluoropropylene (HFP) and vinylidene fluoride (VF2). It was developed by the DuPont Company in 1957 in response to high-performance sealing needs in the aerospace industry. To provide even greater thermal stability and solvent resistance, fluoroelastomer terpolymers containing TFE were introduced in 1959 and in the mid- to late 1960s lower viscosity versions of FKMs were introduced as well. A breakthrough in cross-linking occurred with the introduction of the bisphenol cure system in the 1970s. Bisphenol cure system offered much improved heat and compression set resistance with better scorch safety and faster cure speed. In the late 1970s and early 1980s fluoroelastomers with improved low-temperature flexibility were introduced by using perfluoromethylvinyl ether (PMVE) in place of HFP. These polymers require a peroxide cure. The latest FKM generation shows a much broader fluids resistance profile than standard fluoroelastomers and are able to withstand strong bases and ketones as well as aromatic hydrocarbons, oils, acids and steam.

Currently, fluoroelastomers are generally made by emulsion polymerisation. Fluoromonomers such as HFP, VF2 and TFE are fed into a reactor under elevated temperature and pressure along with surfactants and other additives. Once the polymerisation is completed the latex is removed, coagulated and washed; then, the polymer is dried and packaged for use.

FKMs are characterised by good resistance to compression set, even at high temperatures, to atmospheric oxidation, sun, heat, ozone and weather, and to fungus and mould. Again, they are inherently more resistant to burning than other non-fluorinated hydrocarbon rubbers. Besides their excellent resistance to degradation by a greater variety of fluids, solvents and chemicals than any other elastomer, fluoroelastomers show an extremely low permeability to a broad range of substances.

Compared to most other elastomers, FKM is better able to withstand high temperature, while simultaneously retaining good mechanical properties. Oil and chemical resistances are also essentially unaffected by elevated temperatures. Thermal stability varies within the different fluoroelastomers. The VF2-containing elastomers are slightly less stable due to thermally induced dehydrofluorination. These elastomers should not be used in a totally confined environment.

The outstanding heat stability and excellent oil resistance of FKM are essentially due to the high ratio of fluorine to hydrogen, the strength of the C–F bond and the absence of unsaturation. Generally speaking, increasing the fluorine content, the resistance to fluids and to the chemical attack is improved while the low-temperature characteristics are diminished. There are, however, a few specialty-grade fluorocarbons that can provide high fluorine content with low-temperature properties.

Fluoroelastomers in general show low-temperature properties dictated by two factors: the size of the fluorine atom and the various intermolecular forces that come into play due to high electronegativity of fluorine.

4.5.1 Fluoroelastomer classification

Generically, fluoroelastomers are referred to as FKM polymers following the nomenclature noted in ASTM D1418 [6]. This standard lists three categories 'FKM, FFKM and FEPM' based on the monomers used in the polymerisation process, and gives their general descriptions and the ultimate service applications.

FKM is the largest category of fluoroelastomers covering more than 80% of the whole market. Its D1418 description is the following: 'FKM is a fluoro rubber of the polymethylene type that utilises vinylidene fluoride as a comonomer and have substituent fluoro, alkyl, perfluoroalkyl or perfluoroalkoxy groups in the polymer chain, with or without a cure-site monomer' [6].

FKM can be classified based on the monomer composition and ratio or on the curing chemistry: bisphenol and peroxide. There are currently five logical FKM categories based on the polymerisation of a limited number of monomers:

- *Type 1*. Copolymers of HFP and VF2 have general purposes and are the best balance of overall properties. Fluorine content: 66 wt%. Curing: bisphenol.
  $CH_2 = CF_2$ VF2
  $CF_2 = CF–CF_3$ HFP

- *Type 2.* Terpolymers of TFE, HFP and VF2 show higher heat resistance and have the best aromatic solvent resistance with respect to Type 1. Fluorine content: 68–69.5 wt%. Curing: bisphenol or peroxide.

  $CF_2 = CF_2$ TFE
- *Type 3.* Terpolymers of TFE, a fluorinated vinyl ether and VF2 have improved low-temperature performance and higher cost. Fluorine content: 62–68 wt%. Curing: bisphenol or peroxide.

  $CF_2 = CF–O–CF_3$ perfluoroalkyl vinylether (PAVE)
- *Type 4.* Terpolymers of TFE, propylene and VF2 show improved base resistance, higher swelling in hydrocarbons and decreased low-temperature performance. Fluorine content: 67 wt%. Curing: bisphenol.
- *Type 5.* Pentapolymers of TFE, HFP, ethylene, a fluorinated vinyl ether and VF2, with improved base resistance, low swelling in hydrocarbons and improved low-temperature performance. Fluorine content: 69 wt%. Curing: peroxide.

The standard FFKM designation states that they are 'perfluoro rubbers of the poly-methylene type having all substituent groups on the polymer chain either fluoro, perfluoroalkyl, or perfluoroalkoxy groups' [6]. Usually, perfluoroelastomers are TFE/PAVE copolymers.

The third category, FEPM, is thus defined: 'FEPM is a fluoro rubber of the poly-methylene type containing one or more of the monomeric alkyl, perfluoroalkyl, and/or perfluoroalkoxy groups with or without a cure-site monomer (having a reac-tive pendant group)' [6].

4.5.2 Curing systems for fluoroelastomer

In addition to the inherent differences between the various types and families of fluoroelastomers, a number of compounding variables have major effects on the characteristics of the final vulcanisates. A very important one is the curing system.

In order to cure fluoroelastomers, there are two distinct halogen elimination reactions that are used to develop cross-linking in hydrofluorocarbon elastomers:

- *E1 mechanism:* the cross-linking mechanism is the ionisation provided by an electrophile (peroxide radical). It normally operates without a base. The specific location is a cure-site monomer having an iodine or bromine substitution that is readily displaced by the peroxide radical. In the case of perfluoroelastomers, there are several suitable cure-site monomers having a reactive pendant group, thus leaving the backbone intact and allowing to improve the long-term heat resistance.
- *E2 mechanism:* the simultaneous departure of hydrogen and the adjacent fluo-rine is initiated by a nucleophile (base). This is the logical route reaction for

creating the cross-link site (a double bond in the backbone) for the VF2-containing elastomers.

After developing the cross-link site, curing may follow different reaction pathways:
- Addition → diamine cross-linking
- Aromatic nucleophilic substitution → dihydroxy cross-linking (bisphenol curing)
- ENE reaction → triazine cross-linking (peroxide curing)

The first two mechanisms are ionic, whereas the third one is free radical. A comparison of the various processing and physical properties of FKM compounds synthesised using the three cure systems is shown in Tab. 4.3.

Tab. 4.3: Comparison of cure systems used in cross-linking FKM.

Property and processing characteristic	Types of cure system		
	Diamine	Bisphenol	Peroxide
Processing safety (scorch)	P–F	E	E
Fast cure rate	P–F	E	E
Mould release/mould fouling	P	G–E	G–E
Adhesion to metal inserts	E	G	G
Compression set resistance	P	E	E
Steam, water, acid resistance	F	G	E
Flex fatigue resistance	G	G	G

Rating: E, excellent; G, good; F, fair; P, poor.

Even if diamine curatives (introduced in 1957) have relatively slow curing and do not provide the best possible resistance to compression set, they do offer unique advantages. For example, compounds cured with them exhibit excellent adhesion to metal inserts and high hot tensile strength. Chemically, any amine initiates dehydrofluorination at a VF2 site, that is, followed by the amine addition (cross-link). The subsequent hydrogen fluoride molecule reacts with magnesium oxide (a compound ingredient) rearranging to form magnesium fluoride and water. The water is loosely bound as magnesium hydroxide. Thus, the formation of every cross-link is accompanied by 1 mol of water and 1 mol of carbon dioxide. This is essentially an equilibrium reaction until the water is driven off by an extended high-temperature post-cure. Currently, the amine cure system still enjoys some popularity as it enhances rubber-to-metal bonding.

Most fluoroelastomers are cross-linked with bisphenol AF, introduced in the early 1970s, in the first commercial curative-containing precompound. Bisphenol cured FKMs exhibit fast rates of cure, excellent scorch safety, heat and hydrolytic stability and resistance to compression set. Also in this case the vinylide fluoride presence is necessary to develop the cross-linking site. In 1987, an improved bisphenol curative was introduced, providing faster cure rates, improved mould release and slightly better resistance to compression set, compared to the original bisphenol cure systems.

In 1976, the efficient peroxide curing of fluoroelastomers was made possible for the first time by Dupont for an improved low-temperature fluoroelastomer based on VF2 and perfluoromethyl vinyl ether. Peroxide cross-linking occurs at specific sites available on the cure-site monomer. The very acidic VF2 and the ionic cure mechanisms create the cleavage of the trifluoroalkoxy group on the PMVE resulting in the backbone cleavage, and in the formation of perfluoromethanol as by-product. The addition of the cure-site monomer allows an orderly cross-linking (electrophilic as opposed to nucleophilic) process whereby the triazine structure becomes the cross-link.

Peroxide cure systems provide fast cure rates and excellent physical properties in polymers, which cannot be readily cured with either diamine or bisphenol. Again, the peroxide cure provides enhanced resistance to aggressive automotive lubricating oils, steam and acids. Generally, fluoroelastomer vulcanisates cured with peroxide show higher chemical resistance, but lower thermal properties when compared to the same polymers cured with bisphenol.

4.5.3 FFKM

FFKM elastomers are the highest performing elastomers with excellent thermal and chemical resistance. It was originally introduced by Dupont under the tradename Kalrez® in the 1970s. They are correctly defined as perfluoroelastomers (totally fluorinated, no hydrogens) and are basically TFE/PAVE copolymers, including the cure-site monomer.

FFKMs have outstanding fluid and chemical resistance, dielectric properties and high temperature resistance (up to 327 °C). They show very low volume swell in aggressive chemicals and plasmas and long seal life for low operating costs, so their major market (80%) is essentially in solvent-resistant applications. Their advanced properties bring unrivalled performance to seal integrity, helping to reduce the maintenance and operating costs and improving safety. Specifically, they are used in chemical and hydrocarbon processing, semiconductor manufacturing, aerospace engines, FDA-compliant food, pharmaceuticals and beverages.

4.6 Ethylene–propylene rubber

Nowadays, EPDM rubber, introduced in the market in the early 1960s, is one of the fastest growing polymers because of its certain unique properties [11, 12]. The classification of EPDM includes commercial grades with an ethylene content from 40 to 80 wt%. Generally, ethylene is in a greater amount with respect to propylene.

In the standard binary copolymers, EPMs combine to form a chemically saturated stable polymer backbone providing excellent heat, oxidation, ozone and weather aging resistance. A third, non-conjugated diene monomer can be terpolymerised in a controlled manner to maintain a saturated backbone, thus obtaining the unsaturated ternary copolymers EPDM. The diene is introduced in small quantity (3–8 wt%) to supply the reactive sites necessary to chemically tie all the macromolecular chains together during the vulcanisation process.

In the past, dicyclopentadiene (DCPD) was mostly used as a third monomer, but the corresponding rubbers were slow curing. The recent trend moves towards faster curing grades, using 1,4-hexadiene (HD) and ethylidene norbornene (ENB). These monomers contain two double bonds, one of which is consumed during the polymerisation, while the other remains in the resulting polymer. Each kind of diene incorporates in the main elastomer backbone with a different tendency for introducing long-chain branching or polymer side chains that influence both processing and rates of vulcanisation by sulphur or peroxide cures (i.e., dicumyl peroxide). These characteristics are summarised in Tab. 4.4.

Tab. 4.4: Influence of diene termonomers on the EPDM characteristics.

Termonomer	Cure and property features	Long-chain branching
ENB	Fastest and highest state of cure Good tensile Good compression set resistance	Low to moderate
HD	Intermediate cure rate High scorch safety	Moderate
DCPD	Slow sulphur cure Good compression set resistance	High

EPDM rubbers differ significantly from the diene hydrocarbon rubbers. Since the level of unsaturation is much lower, EPDM has a higher heat and oxygen resistance. Heat aging resistance up to 130 °C can be obtained with properly selected sulphur acceleration systems; on the other hand, heat resistance up to 160 °C can be reached by peroxide curing. As the reactive sites are pendant (not part of the polymer backbone), there are no weak points on the main chains. This means that the reactive entities in the atmosphere (e.g., ozone, which is very aggressive) cannot attack and degrade EPDM, that is, up to 1,000 times more resistant to ozone than NR.

Polymerisation and catalyst technologies in use today allow to design ethylene–propylene polymers to meet specific and demanding applications and processing needs that are important to fit the ever-increasing demands of product quality, uniformity and performance. Since their introduction, ethylene–propylene elastomers have been synthesised using the Zeigler–Natta catalysts in order to polymerise the monomers into controlled polymer structures. Most recently, metallocene catalysts are in commercial use.

Ethylene–propylene elastomers can be produced ranging from amorphous to semi-crystalline structures depending on the polymer composition and how the monomers are combined. As the ethylene content increases, crystallinity increases. Semi-crystalline grades have an ethylene content higher than 60 wt%. When the propylene content increases, EPDM becomes amorphous, showing an elastic behaviour and lower hardness.

EPDM has good resistance to vegetable and hydraulic oils, but very poor resistance to mineral oils and diester-based lubricants. As non-polar elastomers, EPDM shows good electrical resistivity, as well as resistance to polar solvents, such as water, acids, alkalis, phosphate esters and many ketones and alcohols. These elastomers provide excellent resistance to UV light (colour stability) and good low-temperature flexibility properties. Amorphous or low crystalline grades have excellent low-temperature flexibility with glass transition temperatures around −60 °C. Compression set resistance is good, particularly at high temperatures, if sulphur donor or peroxide cure systems are used.

EPDM compounds generally carry high loading of oils such as paraffinic and naphthenic oils without too much loss in the vulcanisate properties. In order to get good properties, the use of reinforcing black or white filler is recommended. In addition, these rubbers can develop high tensile and tear properties, excellent abrasion resistance as well as improved oil swell resistance and flame retardance.

The excellent resistance of EPDM to heat, oxidation, ozone and weathering up to 150 °C, due to the stable, saturated polymer backbone structure, is expected to provide continued value in demanding automotive, construction and mechanical goods applications. Versatility in polymer design and performance has resulted in broad use in automotive weather-stripping and seals, glass-run channel, radiator, garden and appliance hose, tubing, belts, electrical insulation, roofing membrane, rubber mechanical goods, plastic impact modification, thermoplastic vulcanisates and motor oil additive applications. EPDM compounds are not recommended for gasoline, petroleum oil and greases, and hydrocarbon solvent environments, but they are very effective for outdoor applications requiring long-term weathering properties. EPDM elastomers are also suitable for use in hot water and steam environments and for high-temperature brake fluid applications.

4.7 Polybutadiene (BR)

During World War I, polybutadiene was prepared by metallic sodium-catalysed polymerisation of butadiene as a substitute for NR [9–11]. However, polymers prepared by this method, and later by free radical emulsion polymerisation technique, did not possess the desirable properties for the application as a useful rubber. With the development of the Ziegler–Natta catalysts in the 1950s, it has been possible to produce polymers with the controlled stereoregularity needed for having an elastomer.

The most distinguishing feature of polybutadiene is its microstructure, which comprises the ratio of cis, trans and vinyl configuration. Polymers containing 90–98% of a *cis*-1,4-structure can be produced by solution polymerisation using Zeigler–Natta or metallocene catalyst systems. Therefore, the selection of both catalyst and processing conditions allows to synthesise polybutadienes with various distributions of each isomer (microstructure), and with different levels of chain linearity, branching, molecular weight and molecular weight distribution. Solution polymers are characterised by fairly narrow molecular weight distribution and less branching than emulsion butadiene, which account for some of the major differences in processing and performance.

The structure of *cis*-1,4-polybutadiene is very similar to that of NR. Both the polymers are unsaturated hydrocarbons, whereas in the NR the double bond is activated by the presence of a methyl group, the polybutadiene molecule, which contains no such group is generally less reactive. Since the methyl side group tends to stiffen the polymer chain, the glass transition temperature of polybutadiene is −100 °C, less than that of NR molecule. This lower T_g has a number of effects on the final properties of polybutadiene. For example, at room temperature, polybutadiene compounds generally have higher resilience than NR. In turn, this means that the polybutadiene rubbers have a lower heat build-up and this is important in tyre application.

On the other hand, PB shows poor tear resistance, tack and tensile strength. For this reason, these rubbers are seldom used on their own but more commonly in conjunction with other elastomers. For example, they are blended with NR in the manufacture of truck tyres and with SBR in the manufacture of passenger car tyres, taking advantages of its inherently good hysteresis properties, improved durability, abrasion resistance and crack growth resistance. PB generally should not be blended with polar elastomers such as NBR.

Polybutadiene-based compounds can be sulphur cured with systems activated by zinc oxide and fatty acids or cross-linked by peroxides, especially high *cis*-BRs.

Significant amount of polybutadiene is used in footwear and belting compounds as a means of improving abrasion and durability. The outstanding resilience or abrasion resistance has been utilised in the manufacture of solid golf balls and high rebound toy balls and shock absorber. High polybutadiene as well as butadiene–styrene rubbers are used extensively as a modifier of styrene to make high impact polystyrene.

4.8 Styrene–butadiene rubber

SBR is the highest volume and the most important general-purpose synthetic rubber worldwide [11, 12]. Known as Buna-S, it was originally developed to replace NR in tyres. Although it was of poor quality with respect to NR, it has achieved a high market penetration, thanks to the low cost and the higher level of product uniformity.

Currently, its largest application is in passenger car tyres, particularly in tread compounds for superior traction and treadwear, thanks to its good traction properties and abrasion resistance. SBR is also used in footwear, foamed products, wire and cable jacketing, belting, hoses and mechanical goods. For many uses, blends of SBR and other rubber, such as NR or *cis*-polybutadiene, are made.

SBR is an unsaturated copolymer of styrene and 1,3-butadiene. With the exception of some special grades, typically the styrene content is around 23 wt%. Monomers are randomly arranged in the chain.

Several SBRs produced by solution processes are offered commercially. Solution SBRs can be tailored in polymer structure and properties to a much greater degree than their emulsion counterparts. The solution copolymers have narrower molecular weight distribution, less chain branching, higher cis content, lighter colour and less non-rubber constituents than the emulsion SBRs. As a result, they have better abrasion resistance, better flexibility, higher resilience and lower heat build-up than the emulsion rubber. Tensile, modulus, elongation and cost are comparable.

SBR made in emulsion usually contains about 23% styrene randomly dispersed with butadiene in the polymer chains. SBR made in solution contains about the same amount of styrene, but both random and block copolymers can be made. Both emulsion and solution SBRs are offered in oil-extended versions. These have up to 50% petroleum base oil on polymer weight incorporated within the polymer network. Oil extension of SBR improves the processing characteristics, primarily allowing easier mixing, without scarifying the physical properties.

All SBRs, because of their lower unsaturation, have slower curing than NR and require more acceleration. Zinc stearate (or zinc oxide plus stearic acid) is the most common activator for SBR.

The processing behaviour of SBR is not as good as that of NR. Mill mixing is generally more difficult; it has inferior mechanical properties in the unvulcanised state and it does not exhibit the natural tack, which is essential in plying together or otherwise assembling pieces of unvulcanised rubber. Whereas NR is crystalline, SBR is amorphous due to its molecular irregularity. NR crystallises under tensile stress at ambient temperature showing a good tensile strength. Vulcanisates of SBR on the other hand are weak and it is essential to use reinforcing fillers such as fine carbon blacks to obtain high tensile strength and tear resistance. Black reinforced SBR compounds exhibit very good abrasion resistance, superior to the corresponding black reinforced NR vulcanisates. Due to its lower unsaturation, SBR also has better heat resistance and better heat aging qualities with respect to NR. SBR exhibits very good flex fatigue resistance and it is resistant to many polar-type chemicals such as alcohols and ketones. However, it is not resistant to petroleum-based fluids.

4.9 NR and IR polyisoprene

NR and polyisoprene share the same monomer chemistry [8, 13]. Isoprene is the building block of both rubbers and it can polymerise in four different configurations leading to the peculiar behaviour of these materials (see Section 4.9.2).

Generally speaking, polyisoprenes, both natural and synthetic, are noted for their outstanding resilience, resistance to tear, flex fatigue and abrasion and excellent elasticity. Polyisoprenes also have excellent tensile strength characteristics and can be used in low-temperature environments (around −50 °C). They are not recommended for high heat, ozone, sunlight, petroleum or hydrocarbon environments.

The two isoprenes differ slightly. The purity of synthetic polyisoprene provides more consistent dynamic properties with better weather resistance. Again, synthetic polyisoprene gives a relatively odourless rubber. NR, when compared to the synthetic one, provides slightly better properties in tensile strength, tear resistance, compression set and flexural fatigue resistance.

4.9.1 Natural rubber

NR is the mother of all elastomers: it is the only non-synthetic elastomer in use. The NR presently used by industry is obtained by tapping the latex from the *Hevea brasiliensis* tree. By the end of eighteenth century, the properties of rubber as obtained from the *Hevea* tree, available at that time entirely in the forest of Amazon valley, were known throughout Europe. Europeans found that by systematically tapping the tree, the latex could be extracted regularly. With the development of plantation in the Far East, it was discovered that latex could be preserved by adding ammonia immediately after its collection. This marked the beginning of the commercial latex technology. Presently, apart from Brazil, vast plantations exist in India, Malaysia, Indonesia, Sri Lanka, Vietnam, Cambodia and Liberia.

Tapping is usually done by shaving about 1 or 2 mm thickness of the bark with each cut, usually in the early morning hours, after which latex flows for several hours and it is collected in cups mounted on each tree. The cut is made with special knife or gouge, sloping from left to right at about 20–30° from the horizontal. The content of each latex cup is transferred into bigger containers and transported to storage tanks at the bulking station. NR is produced from the latex in a series of steps involving preservation, concentration, coagulation, dewatering, drying and cleaning. After those, the concentrated latex is blended with various additives depending on the final applications (latex compounding).

The milky liquid exuded by the *Hevea* tree is a colloidal solution of rubber and non-rubber particles in water, having a diameter range between 0.05 and 5 μm. The total solids of fresh field latex vary typically from 30 to 40 wt% depending on weather, age of the tree, method and frequency of tapping, tapping frequency and other factors.

Because of its natural derivation, NR is sold in a variety of grades based on purity (colour and presence of extraneous matter), viscosity, oxidation resistance and the curing rate. Modified NRs are also available, including:

- Epoxidised NR
- Deproteinised NR
- NR in which process oils have been incorporated
- NR with grafted poly(methyl methacrylate) side chains
- Thermoplastic NR: blends of NR and polypropylene

The best types and grades of NR contain at least 90% of *cis*-1,4-polyisoprene together with resins, proteins, sugars and so on. The raw material on the market comprises a molecular weight in the range from 500,000 to 10,000,000, which is very high for processing. Hence, rubber has to be extensively masticated on a mill or in an internal mixer to break down the macromolecules to a size that enables them to flow without any difficulty during processing by extrusion or by moulding. The breakdown occurs more rapidly at either high (120–140 °C) or moderately low (30°–50 °C) temperature. The breakdown at higher temperatures is due to oxidative scission, whereas at lower temperatures it is due to the mechanical breaking of primary bonds. The free radicals thus produced get stabilised by addition of oxygen.

The NR microstructure, containing mostly *cis*-1,4-polyisoprene, determines a T_g of approximatively –75 °C. For synthetic polyisoprene the glass transition temperature is slightly higher (about –70 °C) because of the presence of *trans*-1,2 and -3,4 configurations as well.

Due to its high structural regularity, NR tends to crystallise spontaneously at low temperatures or when it is stretched. Low-temperature crystallisation causes stiffening, which can be easily reversed by warming. Crystallisation gives to the NR high tensile strength and resistance to cutting, tearing and abrasion.

NR is generally vulcanised using accelerated sulphur system in the presence of an activator, such as zinc oxide, and an accelerator. Peroxides are also occasionally used.

NR has a poor resistance to ozone, high temperatures, weathering, oxidation, oils and concentrated acids and bases. Therefore, to achieve the protection against oxidation, non-staining antioxidants, such as hindered phenols, must be used. Where staining can be tolerated, amine derivatives such as phenylene diamines, phenyl beta-naphthylamine and ketone-amine condensates may be added. These have good heat stability and are also effective against copper contamination, which causes the rapid degradation of NR.

The outstanding strength of NR has maintained its position as the preferred material in many engineering applications. It has a long fatigue life, good creep and stress relaxation resistance and low cost. Other than for thin sections, it can be used until approximately 100 °C and sometimes above. It can maintain flexibility until –60 °C, if compounded for the purpose.

The low hysteresis and its natural tack make NR ideal for use in tyre building. NR is used in the carcass of cross-ply tyres for its building tack, ply adhesion and good tear resistance. It is also used in the sidewalls of radial ply tyres for its fatigue resistance and low heat build-up. In tyres of commercial and industrial vehicles, NR content increases with tyre size. Almost 100% NR is used in the large truck and earthmover tyres, which require low heat build-up and maximum cut resistance.

NR is also used in industrial goods, such as hoses, conveyor belts, rubberised fabrics and footwear soles, engineering products, resilient load bearing and shock or vibration absorption components, and latex-based products such as gloves and adhesives.

4.9.2 Polyisoprene (IR)

Polyisoprene is made by solution polymerisation of isoprene (2-methyl-1,3-butadiene) that can polymerise in different isomeric forms: *trans*-1,4-addition, *cis*-1,4-addition, 1,2-addition, leaving a pendant vinyl group, and 3,4-addition.

The successful synthesis of stereoregular polyisoprene fulfilled a goal sought by polymer chemists for nearly a century. Researchers knew that isoprene was the building block for NR, and through the years many attempts were made to synthesise materials with similar properties. Initially, the resulting polymers failed to exhibit some of the desired characteristics of NR because of differences in microstructure, which plays an important role in polyisoprene physical properties. Indeed, the polymer chains in the early synthetics obtained by free radical polymerisation contained mixtures of all possible molecular configurations joined together in a random fashion. Specifically, they lacked the very high *cis*-1,4-structure of the NR backbone that gives the ability to undergo strain crystallisation, which determine the typical physical properties such as green strength, tear and tensile strength.

In the mid-1950s, researchers discovered and developed the Ziegler–Natta catalyst systems that could selectively join together monomer units in a well-ordered fashion. Shortly after Karl Ziegler's breakthroughs in catalyst systems for the polymerisation of ethylene, similar catalysts were developed for isoprene. These stereospecific catalysts allowed the realisation of a nearly pure *cis*-1,4-structure, and thus the production of a synthetic NR.

The initial commercialisation of a stereoregular low *cis*-1,4-IR (92 wt%) was realised in 1960 by the Shell Chemical Company with the introduction of the Shell Isoprene Rubber, produced with an alkyl lithium catalyst (Li-IR). However, the *cis*-1,4-content of Li-IR was insufficient to achieve the important crystallisation properties of NR. In 1962, Goodyear introduced NATSYN®, a Ziegler–Natta (titanium–aluminium)-catalysed IR (Ti-IR) with a *cis*-1,4-content of 98.5%, finally allowing the benefits of strain crystallisation to be realised.

Polyisoprene compounds, like those of NR, exhibit good building tack, high tensile strength, good hysteresis and good hot tensile and hot tear strength. However, the very specific nature of synthetic polyisoprene due to its controlled synthesis provides a number of factors that differentiate it from NR. Polyisoprene is chemically purer (up to 99% of polymer), since it does not contain proteins and fatty acids of its natural counterpart where the hydrocarbon is on average around 93%. Since its molecular weight is lower than that of NR, polyisoprene is easier to process and gives a less variable (although generally slower) cure. It is more compatible in blends with EPDM and SBR, and provides less green strength (pre-cure) than NR. Polyisoprene is added to SBR compounds to improve tear strength, tensile strength and resilience while decreasing heat build-up. Blends of polyisoprene and fast-curing EPDM combine high ozone resistance with good tack and cured adhesion that is uncharacteristic of EPDM alone.

Synthetic polyisoprene compounds having the same plasticity of NR show less die swell because of the lower nerve. Also, at the same plasticity, the synthetic polymer shows significantly faster extrusion rates. Due to the easier processability of IR, less mechanical work and breakdown are required. Shorter mix cycles and the elimination of premixing are thus possible when IR is used as a direct replacement for NR, resulting in time and power saving.

Currently, synthetic polyisoprene is used in a wide variety of industries and as replacement of NR in those applications requiring consistent cure rates, tight process control or improved extrusion, moulding and calendering.

Since the synthetic elastomer can be produced with the very low level of branching, and relatively narrower molecular weight distribution that contributes to lower the heat build-up, certain grades of polyisoprene are used as alternative to NR in the tread of high service tyres (truck, aircraft and off-road) without sacrificing abrasion resistance, groove cracking, rib tearing, cold flex properties and weathering resistance.

Because of the high purity and the high tensile strength, polyisoprene is widely used also in medical goods and food-contact items. These include baby bottle nipples, milk tubing and hospital sheeting, since with polyisoprene are eliminated the potential dangers coming from the natural proteins and impurities found in NR. They also find major use in footwear and mechanical goods.

4.10 Polychloroprene (neoprene)

Neoprene is the trade name for chloroprene polymers (2-chloro-1,3-butadiene) manufactured since 1931 by DuPont de Nemours [11, 12]. Today this material is among the leading special-purpose rubbers with an annual consumption of nearly 300,000 tonnes worldwide due to its favourable combination of technical properties.

In principle, it is possible to polymerise chloroprene by anionic, cationic and Ziegler–Natta catalysis techniques. However, because of product properties and economic considerations, free radical emulsion polymerisation is used exclusively today in a commercial scale, using both batch and continuous processes at 40 °C.

The chloroprene monomer can polymerise in four isomeric forms: 1,4-*trans*, 1,4-*cis*, 3,4- and 1,2-polychloroprene. Neoprene is typically constituted by 88–92% 1,4-*trans*, with a degree of polymer crystallinity proportional to the trans content. 1,4-*cis*-polychloroprene accounts for 7–12% of the structure and 3,4-addition makes up about 1%. The approximately 1.5% of 1,2-addition provides the principal sites of vulcanisation since in this arrangement the chlorine atom is both tertiary and allylic. Accordingly, it is strongly activated and thus becomes a curing site on the polymer chain.

The high amount of *trans*-1,4-units in neoprene leads to a rubber able to crystallise. Indeed, the high structural regularity, related to the high trans content, allows the strain-induced crystallisation that results, as for NR, in high tensile strength.

The physical, chemical and rheological properties of commercial polychloroprenes are dependent on the ability to change the molecular structure by varying the polymerisation conditions (e.g., temperature, monomer conversion, type and amount of comonomers, molecular weight modifiers and emulsifiers). However, all neoprene types have common inherent characteristics such as:
- Resistance to degradation from sunlight, ozone and weather
- Good performances when in contact with oils and many chemicals
- Use over a wide range of temperatures
- Outstanding physical toughness
- Better resistance to burning than hydrocarbon rubbers

The structural similarities between the neoprene and the NR molecule are apparent. However, while the methyl group activates the double bond in the polyisoprene molecule, the chlorine atom exerts opposite effect in neoprene. Thus, the polymer is less liable to oxygen and ozone attack. The chlorine atom has two other positive impacts on the polymer properties. First, the polymer shows improved resistance to oil compared with all hydrocarbon rubbers. In addition, thanks to the presence of chlorine, these rubbers have an increased resistance to burning which may further be improved by the use of fire retardants. These features together with a somewhat better heat resistance than the diene hydrocarbon rubbers have resulted in the extensive use of neoprene over many years. Neoprene is less resistant than NR to low-temperature stiffening but can be compounded to give improved low-temperature resistance.

Polychloroprene can be vulcanised by various accelerator systems over a wide temperature range even if neoprene is generally cured with zinc oxide and magnesium oxide, or lead oxide for enhanced water resistance. Its excellent adhesion to metals makes polychloroprene ideal for moulding with metal inserts.

Neoprene for dry rubber applications is available in three different families: G, W and T types. These materials offer a broad range of physical properties and processing conditions. The main fundamental difference between the G family and the others arises from the fact that the G type is obtained by copolymerisation of chloroprene with sulphur and stabilisation with thiuram sulphide. W and T types do not contain sulphur.

Solid neoprenes are classified as general-purpose, adhesive or specialty types. General-purpose types are used in a variety of elastomeric applications, particularly moulded and extruded goods, hose, belts, wire and cable, heels and soles, tyres, coated fabrics and gaskets. The adhesive types are adaptable to the manufacture of quick setting and high bond strength adhesives. Specialty types have unique properties such as exceptionally low viscosity, high oil resistance or extreme toughness. These properties make specialty neoprenes useful in unusual applications, for example, crepe soles, prosthetic applications, high solid cements for protective coatings in tanks and turbines. Neoprenes are also available in latex form.

4.11 Butyl rubber (IIR)

Homopolymer from isobutylene has little use as a rubber because of high cold flow (T_g about −73 °C) but the copolymer with isoprene, to introduce the unsaturation required for cross-linking, is a rubber which is widely used in many special applications [8, 11, 12].

Butyl rubber is the common name for the copolymer of isobutylene and 1–3% isoprene produced by cold (−100 °C) cationic solution polymerisation using Friedel–Crafts catalysts such as $AlCl_3$ or BF_3. The purity of isobutylene (95% or more) is important for acquiring high molecular weight.

Most of butyl rubber distinguishing characteristics are the result of its low level of chemical unsaturation. The essentially saturated hydrocarbon backbone of the IIR polymer effectively repels water and polar liquids but shows affinity to aliphatic and some cyclic hydrocarbons. Products of butyl rubber, therefore, are swollen by hydrocarbon solvents and oils, but show resistance to moisture, mineral acids, polar-oxygenated solvents, synthetic hydraulic fluids, vegetable oils and ester-type plasticisers. Good resistance to heat, abrasion, oxygen, ozone and sunlight are dependent upon the low saturation level. The high level of ozone and weathering resistance enables butyls to be used in rubber sheeting for roofs and water management applications and high-quality electrical insulation.

Butyl rubber is highly resistant to the diffusion of gas molecules, thanks to its closely packed structure. The high degree of impermeability to gases (i.e., helium, hydrogen, nitrogen, CO_2) makes butyl almost an exclusive choice for use in inner tubes, air barriers for tubeless tyres, air cushions, pneumatic springs, accumulator bags, air bellows and so on.

The molecular structure of the polyisobutylene chain provides less flexibility and greater delayed elastic response to deformation than most elastomers. This imparts vibration damping, good hot tear strength and shock absorption properties to butyl rubber vulcanisates, mostly used in the automotive sector.

As common with more highly unsaturated rubbers, butyl may be cross-linked with sulphur, thanks to the activation by zinc oxide and organic accelerators. In contrast to the higher unsaturated varieties, however, adequate vulcanisation can be achieved with very active thiuram and dithiocarbamate accelerators. Other less active accelerators such as thiazole derivatives can be used as modifiers to improve the scorch safety. Unfortunately, sulphur-cured butyl rubber has relatively poor thermal stability, and soft under prolonged exposure at temperatures above 150 °C because the low unsaturation prevents the oxidative cross-linking. Curing with phenol–formaldehyde resins provides products with very high heat and aging resistance.

4.11.1 Halobutyl rubber

Chloro- and bromobutyl rubbers are commercially the most important derivatives of butyl rubber after their introduction on the market in the early 1960s.

The halogenation reaction is carried out in hydrocarbon solution using elemental chloride and bromine (equimolar with the enchained isoprene). The halogenation is fast and proceeds mainly by an ionic mechanism. More than one halogen atom per isoprene unit can also be introduced. However, the reaction rates for excess halogens are lower and the reaction is complicated by chain fragmentation.

The introduction of a small amount of chlorine (1.2 wt%) or bromine (around 2 wt%) in the butyl polymer gives rise to rubbers, which can be blended better with general-purpose rubbers due to increased polarity.

The main difference between bromo- and chlorobutyl rubbers is the higher reactivity of the C–Br bond compared to that of C–Cl since halogenation provides more active functionalities to the butyl chains. Chlorobutyl is used when longer scorch time is needed. Bromobutyl is preferable when greater reactivity is requested. For both butyl derivatives, the choice of mineral fillers can affect the cure characteristics. For example, acid clays produce very fast cures that can be slowed by addition of magnesium oxide or other scorch retarders.

Halogenated rubbers are used as innerliners for tubeless tyres, tyre sidewall components and heat-resistant truck inner tubes, hose (steam and automotive), gaskets, conveyor belts, adhesives and sealants, tank linings, tyre curing bags, truck cab mounts, aircraft engine mounts, rail pads, bridge bearing pads, pharmaceutical stoppers and appliance parts, thanks to their resistance to the environmental attack and the low permeability to gases.

References

[1]   Seymour, R., Kirsenbaum, G. High performance polymers. Elsevier, New York, USA, 1986.
[2]   Dick, JS. Rubber technology – Compounding and testing for performance. Hanser, Munich, Germany, 2001.
[3]   Rochow, EG. Silicon and silicones. Springer-Verlag, Berlin, Germany, 1987.
[4]   Hardman, B. Silicones. Encycl. Polym. Sci. Eng. 1989, 15, 204.
[5]   Noll, W. Chemistry and technology of silicones. Academic Press, London, UK, 1968.
[6]   ASTM 1418, Standard Practice for Rubber and Rubber Lattices-Nomenclature.
[7]   Stark, FO., Falendar, JR., Wright, AP. In Comprehensive organometallic chemistry, Eds. G. Wilkinson, F. G. A. Stone, E. W. Abel. Pergamon Press, Oxford, UK, 1982, p. 305.
[8]   Gent, AN. Engineering with rubber: how to design rubber component. Hanser-Verlag, Munich, Germany, 1992.
[9]   The synthetic rubber manual, 13th Ed. International institute of Synthetic rubber producers, Houston, Texas, USA, 1995.
[10] Gargani, L., Bruzzone, M. Advances in elastomers and rubber elasticity. Plenum Press, London, UK, 1986.
[11] Bhowmick, AK., Stephens, HL. Handbook of elastomers, 2nd Ed. Marcel Decker, New York, USA, 2001.
[12] The Vanderbilt rubber handbook. 13th Ed., R.T. Vanderbilt Co., Inc., Norwalk, USA, 1990.
[13] Stern, HJ. Rubber, natural and synthetic. Maclaren, London, UK, 1967.

5 Thermoplastic elastomers

5.1 Introduction

Thermoplastic elastomers (TPEs) have many of the physical properties of rubbers, that is, softness, flexibility and resilience, but in contrast to conventional rubbers, they are processed as thermoplastics [1, 2]. Rubbers must be cross-linked by a thermosetting process to give useful properties: vulcanisation is slow and irreversible and takes place upon heating. With TPEs, on the other hand, the transition from a processable melt to a solid rubber-like object is rapid and reversible and takes place upon cooling. Therefore, TPEs combine the functional performance and properties of thermoset rubbers with the processability of thermoplastics. Because the melt to solid transition is reversible, some properties of TPEs, that is, compression set, solvent resistance and resistance to deformation at high temperatures, are usually not as good as those of the conventional vulcanised rubbers. Applications of TPEs are, therefore, in areas where these characteristics are less important, for example, footwear, wire insulation, adhesives, polymer blending, and not in areas such as tires.

The two most important manufacturing methods with TPEs are extrusion and injection moulding. Compression moulding is seldom, if ever, used. Fabrication via injection moulding is extremely rapid and highly economical. Both equipment and methods normally used for extrusion or injection moulding of a conventional thermoplastic are generally suitable for TPEs. TPEs can also be processed by blow moulding, thermoforming and heat welding.

TPE materials have the potential to be recyclable since they can be moulded, extruded and reused like plastics, whereas rubbers are not recyclable owing to their thermosetting characteristics. TPEs also require little or no compounding, with no need to add reinforcing agents, stabilisers or cure systems. Hence, batch-to-batch variations in weighting and metering components are absent, leading to improved consistency in both the raw materials and the fabricated items. Besides that, TPEs consume less energy in process and a closer and more economical control of product quality is possible.

TPEs can be customised in a number of ways to suit the needs of applications. Modifications include hardness, physical properties, service temperature and bondability. They are used where conventional elastomers cannot provide the required range of physical properties, finding large application in the automotive sector, medical devices, mobile electronics and in household appliances sector.

In addition to their use in stand-alone applications, TPEs can be overmoulded onto a rigid plastic substrate using co-injection or insert injection moulding technology. The result is an outer layer of TPE material over a rigid

https://doi.org/10.1515/9783110640328-005

substrate, contributing a distinct, soft-touch feel to the stiffness and strength of the supporting structure. TPEs can adhere to many different plastic substrates, including polypropylene, polycarbonate (PC), acrylonitrile-butadiene-styrene (ABS), PC/ABS, copolyester, polyamide and polystyrene, for a seamless and permanent bond.

The addition of TPEs is an affordable way to make an ordinary product stand out and to justify a better value to buyers. A few grams of a TPE material can often make a major difference in the perceived value and quality of an item. Further, incorporation of a TPE via over-moulding can eliminate some secondary operations for further cost savings.

Today, even the most basic and traditional consumer products have evolved into sophisticated and highly differentiated and personalised brands. TPEs are playing an important role in this evolution by allowing designers to add ergonomic features, unique sensory benefits and appealing aesthetics.

Versatility and other unique properties of TPEs can be leveraged by designers to deliver superior performances such as surface feel and texture. A wide range of textures can add sensory interest to a product, such as a bubble or ridged surface or a leather-like feel, that can replace expensive secondary painting or adhesion operations. TPEs help to improve not only touch and appearance, but also some functional performances as in the case of cushioning technology for footwear, soft grip handles for toothbrushes and so on.

TPEs are available in a wide array of colours and effects, which can amplify their quality and contribute to the brand identity and shelf appeal. TPEs can be customised with bright colours and effects such as metallics, sparkles and pearlescents and various shine levels, including eye-catching high gloss, which can provide a high-tech or high-fashion aesthetic. They may be produced in opaque, semi-transparent and transparent formulations for additional aesthetic interests. TPEs can even be infused with a scent, such as the material used to create mosquito repellent wristbands, pet tags and hanging grids, which incorporates a proprietary formula of all-natural essential oils and fragrance to repel mosquitoes.

Because of the increased production and lower cost of raw materials, TPEs are a significant and growing part of the total polymers market. Global consumption of thermoplastic rubbers of all types is estimated at about 1,300,000 tons/year with a growth of more than 10% per year [2]. Of this, the market is estimated to be divided as follows: styrenic block copolymers, 50%; blends and thermoplastic vulcanisates based on polyolefins, 29%; polyurethane block copolymers and polyester block copolymers, 16%; and others, 6%.

Several authors analysed the advantages and drawbacks of the use of TPEs in specific fields of application, as Baumann reviewed the innovative applications of TPEs, showing their versatility and advantages in design and economy [3].

5.2 Historical background

The real era of TPEs began with the advent of block and graft copolymers. However, some old blends are tacitly accepted as TPEs, even though their structure does not exhibit some of their essential characteristics, such as the separation between soft and hard phases. In this framework, PVC plasticised by high-boiling liquids is often considered as one of the precursors of the TPEs.

The two-step reaction between diols and diisocyanates resulting in polyurethanes was an important point in the TPE development, since these elastomers exhibit a very rapid elastic recovery and good processability [4, 5]. The first polyurethane samples resulted from the pioneering work of Bayer [6] and Christ [7] aiming at the preparation of new textile fibres. Coffey described their elastomeric properties [8]. However, the really scientific approach to thermoplastic polyurethanes (TPU) began with the publications of Müller et al. [9] and Petersen et al. [10]. The theoretical study by Bayer and his school describes for the first time, a truly linear polyurethane prepared through a sequence of polyaddition steps, announcing the classical route used to obtain thermoplastic polyurethanes: preparation of an α,β-diisocyanate prepolymer, resulting from the reaction of an α,β-dihydroxy-polyester (or polyether) with an excess of diisocyanate, then extended by water with formation of urea linkages and then reacted with additional diisocyanate [6]. When the importance of the addition of a short-length diol was recognised, the modern chemistry of polyurethanes was definitely born.

The subsequent introduction of thermoplastic polyurethanes on the market was the result of a new strategy, rapidly applied also to polyesters when Snyder polycondensed a mixture of terephthalic acid, octanedioic acid and propane-1,3-diol and, separately, terephthalic acid with ethane-1,2-diol [11]. When mixed, these two polycondensates reacted and the ultimate product exhibited both elastomeric and plastic properties.

Both polyurethanes and Snyder's product behaved as vulcanised rubber even though they were not chemically cross-linked, as revealed by their complete solubility. Soon it became clear that the TPE chains were formed by a succession of long, flexible blocks, responsible for the elasticity, and hard blocks, interconnecting the macromolecular chains and acting as reversible physical cross-links. It appeared that the blocks were incompatible and localised in separate microdomains. This microphase separation was called "segregation."

When Szwarc et al. discovered the anionic living polymerisation, a completely different preparation of these elastomers was proposed and the research on TPEs passed from infancy to maturity [12, 13]. They used sodium-metal-naphthalene initiators to prepare poly(styrene-b-isoprene-b-styrene), which can be considered as the first TPE with a perfectly defined structure. However, this copolymer was not commercialised, as most of the polyisoprene units were -3,4-, giving poor elastomeric properties. It is only when the polymerisation was initiated by alkyl lithium that poly(styrene-b-isoprene-b-styrene) and poly(styrene-b-butadiene-b-styrene) with the classical TPE properties (high tensile strength and elongation at break, rapid elastic recovery and no chemical cross-linking) were synthesised. Bailey et al. announced the successful synthesis of these materials in 1966 [14] and Holden et al. published the corresponding theory in 1967, extending it to other block copolymers [15].

Besides their commercial success, TPEs were the result of logical considerations and scientific efforts, giving birth to a new field of science and technology. These multiphase materials stimulated many theoretical and experimental studies dealing not only with their chemistry and synthesis, but also with their structure and morphology, and consequently with their characterisation. The understanding of the physical cross-linking was perhaps more important than the chemistry. Thus, these materials were first called virtually cross-linked elastomers [16].

Based on these fundamental studies, many other TPEs were prepared following different synthetic processes, resulting both in commercial or not materials. In 1962, this new strategy was applied to prepare copolymers containing random poly (ethylene-co-propylene) as amorphous blocks, and linear polyethylene or isotactic polypropylene as hard blocks [17]. Once more, Tobolsky predicted that such a copolymer would exhibit TPE properties [18].

In 1970, Hartman and coworkers reported on a series of elastomeric thermoplastics prepared by grafting butyl rubber onto polyethylene [19]. Butyl rubber is beneficial in improving the environmental stress-cracking resistance and impact resistance of PE at normal and low temperatures. Addition of substantial amounts of polyethylene to butyl rubber resulted in excellent resistance to degradation by water and high temperature [20]. The above fundamental studies were followed by the application of polycondensation mechanism that resulted in the preparation of some very important TPEs such as poly(amide-b-ester) [21], poly(amide-b-ether) [22] and poly(ether-b-ester) [23].

Nowadays, the development of TPEs concerns many branches of macromolecular chemistry: cationic and radical polymerisations, chemical modification, enzymatic catalysis or the use of microorganisms. The elastomers based on halogen-containing polyolefins and those prepared by dynamic vulcanisation are also included in the TPE family.

Over recent years, blends of natural rubber with polyolefins have been studied as well [24]. They are called thermoplastic natural rubbers (TPNR). Michaeli et al. investigated the mechanical properties and morphology of NR/LLDPE blends, founding that the ozone resistance of the TPE was very high [25].

5.3 Thermoplastic elastomers: key characteristics

TPEs sometimes are referred as thermoplastic rubbers, since they show both thermoplastic and elastomeric properties [26]. The principal difference between thermoset elastomers and TPEs is the type of cross-linking in their structures that are chemical in the first case (irreversible) and physical in the second one (reversible). Therefore, a TPE has all the same features of an elastomer except that the chemical cross-linking is replaced by a network of physical cross-links that do not exist permanently and may disappear with the increase of temperature.

The ability to form physical cross-links is linked to well-defined structural requirements. TPEs must be two-phase materials, and each molecule must consist of two opposite types of structure, one corresponding to the elastomeric part and the second one consisting in the hard part that forms the physical cross-links. In other words, TPEs are known as two-phase systems consisting in a continuous phase (also called domain) that exhibits an elastic behaviour (soft phase) and a dispersed phase that represents the physical cross-links (hard phase). If the dispersed phase is elastic, the polymer is a toughened thermoplastic, not an elastomer. Because of this structure TPEs have similar characteristics to rubber materials like flexibility, elasticity and resistance to weather conditions, due to the soft phase. Due to the hard phase they also have similar characteristics to thermoplastic material such as heat resistance, resistance and ability to be processed.

In order to qualify a TPE three essential characteristics have to be fitted:
1. Ability to be stretched to moderate elongations and, upon the removal of stress, to return to its original shape
2. Processing as a melt at elevated temperature
3. Absence of significant creep

TPEs are primarily divided into two main groups that may be separated in different classes according to their structures. The first group consists of block copolymers where the soft and hard phase belong to the same macromolecular chain, such as styrenic block copolymers, polyamide/elastomer block copolymers (COPAs), polyether ester block copolymers (COPEs) and polyurethane block copolymers (TPUs). The simplest structure is an A–B–A block copolymer, where A is the hard phase and B the elastomeric one, as in the case of poly(styrene-b-butadiene-b-styrene). The hard phase at room temperature becomes fluid upon heating, whereas the soft phase is rubber-like at room temperature.

Polystyrene is a glassy polymer with glass-transition temperature (T_g) around 100 °C so it will resist to flow and creep at ambient temperatures, but it can flow and be moulded at temperatures above T_g. 1,4 cis polybutadiene needs to be cross-linked to be a useful elastomer. The styrene-butadiene diblock copolymers possess a two-phase microstructure due to incompatibility (not miscibility) between the polystyrene and polybutadiene blocks, the former separating into domains having sphere or rod shapes depending on the exact composition (Fig. 5.1). These domains are attached to the ends of elastomeric chains and form multifunctional junction points, thereby providing the physical cross-links to the rubber.

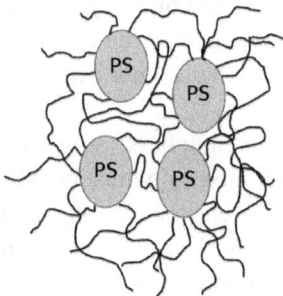

Fig. 5.1: Schematic microstructure of a styrene block copolymer (the PB chains are shown in black).

The butadienic matrix phase gives the overall elastomeric response while the dispersed styrenic islands are the restraining physical cross-links. When they are heated, the polystyrene domains soften and the copolymer becomes processable as thermoplastics. When cooled, it exhibits good elastomeric properties as the polystyrene domains reform and strength returns. With low polystyrene content, the material is elastomeric with the properties of polybutadiene predominating.

This explanation of the properties of TPEs has been given in terms of a poly(styrene-b-elastomer-b-styrene) block copolymer, where the elastomer is butadiene, but it would apply to any block copolymer with the multi-block structure A–B–A–B, as well as to branched block copolymers (A–B)$_{nx}$ (where x represents an n-functional junction point). In principle, A can be any polymer normally regarded as a hard thermoplastic, and B can be any polymer normally regarded as elastomeric, as shown in Tab. 5.1.

Block copolymers can microphase separate to form periodic nanostructures, as well highlighted in the styrene-butadiene-styrene block copolymer, known as Kraton, and used for shoe soles and adhesives. Transmission electron microscopy (TEM) is one of the most powerful equipment to characterize the structure and morphology of this kind of TPEs. Sometimes, TEM is associated with other techniques, particularly small-angle X-ray scattering (SAXS), wide-angle X-ray scattering (WAXS), small-angle neutron scattering (SANS) and atomic force microscopy (AFM), to provide more information on the microstructure of TPEs. For example, the spacing between domains may be evaluated by SAXS.

The second group of TPEs comprises the elastomer blends formed by mixing soft and hard phases, as shown in Tab. 5.2. The production of the hard polymer/elastomer combinations, under conditions of intensive shear, is simpler than that of block copolymers. To achieve a satisfactory dispersion, both viscosities and

Tab. 5.1: Thermoplastic elastomers based on block copolymers (T: triblock, A–B–A; B: branched, (A–B)$_n$x; M: multiblock, A–B–A–B···; X: mixed structures, including multi-block).

Hard segment (A)	Soft segment (B)	Structure
Polystyrene	Polybutadiene and polyisoprene	T, B
Polystyrene	Poly(ethylene-*co*-butylene) and poly(ethylene-*co*-propylene)	T
Polystyrene and substituted polystyrenes	Polyisobutylene	T, B
Poly(α-methylstyrene)	Polybutadiene, polyisoprene	T
Poly(α-methylstyrene)	Poly(propylene sulphide)	T
Polystyrene	Polydimethylsiloxane	T, M
Poly(α-methylstyrene)	Polydimethylsiloxane	T
Polysulphone	Polydimethylsiloxane	M
Poly(silphenylene siloxane)	Polydimethylsiloxane	M
Polyurethane	Polyester and polyether	M
Polyester	Polyether	M
Poly(β-hydroxyalkanoates)	Poly(β-hydroxyalkanoates)	M
Polyamide	Polyester and polyether	M
Polycarbonate	Polydimethylsiloxane	M
Polycarbonate	Polyether	M
Polyetherimide	Polydimethylsiloxane	M
Polymethyl methacrylate	Poly(alkyl acrylates)	T, B
Polyurethane	Poly(diacetylenes)	M
Polyethylene	Poly(α-olefins)	M
Polyethylene	Poly(ethylene-*co*-butylene) and poly(ethylene-*co*-propylene)	T
Polypropylene (isotactic)	Poly(α-olefins)	X
Polypropylene (isotactic)	Polypropylene (atactic)	X

solubility parameters of the polymers must be carefully matched at the mixing conditions. In some cases, grafting may occur as well.

In a variation of this blending technique, the elastomer can be cross-linked while the mixing is taking place. This process is defined as *dynamic vulcanisation* and gives inter-dispersed multiphase systems in which there is a finely dispersed and cross-linked elastomer phase (see Section 5.6). Blends of polypropylene with EPDM were the first materials of this family. Blends with ethylene–propylene copolymers are now commercially more important, and propylene copolymers often replace polypropylene homopolymer as the hard phase.

Other TPE blends, in which the elastomer phase may or may not be cross-linked, include blends of polypropylene with nitrile, butyl and natural rubbers; blends of PVC with nitrile rubber and plasticisers and blends of halogenated polyolefins with ethylene interpolymers. They can be melt-reprocessed numerous times and exhibit good resistance to heat, oils and many chemicals. They are single-

Tab. 5.2: Thermoplastic elastomers based on hard polymer/elastomer combinations (B: Simple blend; DV: dynamic vulcanisation).

Hard polymer	Soft polymer	Structure
Polypropylene	EPR or EPDM	B
Polypropylene	EPDM	DV
Polypropylene	Poly(propylene/1-hexene)	B
Polypropylene	Poly(ethylene/vinyl acetate)	B
Polypropylene	Butyl rubber	DV
Polypropylene	Natural rubber	DV
Polyamide	Nitrile rubber	DV
Polypropylene	Nitrile rubber	DV
PVC	Nitrile rubber + DOP (dioctyl phthalate)	B, DV
Halogenated polyolefins	Etheylene interpolymer	B
Polyester	EPDM	B, DV
Polystyrene	SBS + oil	B
Polypropylene	SEBS + oil	B

phase polymers, in contrast to most TPEs, which are two-phase systems, and for this reason they are often defined as *processable rubbers*.

Commercially important TPE combinations based on blends of polystyrene with S–B–S and oil and also on blends of polypropylene with S–EB–S and oil have been investigated as well, as the same blends with silicone oils. Also TPEs based on blends of a silicone rubber, cross-linked during processing with thermoplastic block copolymers, have been produced [27]. Graft copolymers and elastomeric ionomers have not become commercially important.

Summarising, currently on the market six generic classes of commercial TPEs can be identified:

1. Styrenic block copolymers (TPE-s)
2. Dynamically vulcanised blends elastomeric alloys (TPE-v or TPV)
3. Polyolefin blends (TPO)
4. Thermoplastic polyamides (COPA)
5. Thermoplastic copolyester (COPE)
6. Thermoplastic polyurethanes (TPU)

5.4 Synthesis of block copolymers

The synthesis of block copolymers follows two essential pathways: (i) a difunctional oligomer initiates the formation of two or more other blocks and (ii) two or several different difunctional oligomers react together or with a coupling agent. Sometimes, the second block can be prepared in the presence of the first one. The first pathway is mainly encountered in chain polymerisation (anionic, cationic and controlled radical polymerisations); the second one refers essentially to polycondensation and polyaddition.

The most important TPEs prepared by polycondensation are polyester-based TPEs, poly(amide-b-ethers), polyurethanes and so on. However, some less known condensation TPEs are metal-containing macrocycles as monomers, liquid crystalline side chains and metallo-supramolecular block copolymers.

Only three common monomers, styrene, butadiene and isoprene, are easy to polymerise anionically [28, 29]. Therefore, only two useful A–B–A block copolymers, S–B–S and S–I–S, can be produced directly. In both cases, the elastomer segments contain double bonds, which are reactive and limit the stability of the product. Molar mass distribution is characteristically low for anionic polymerisation, so the macromolecular architecture is accurately controlled. To improve stability, the polybutadiene mid-segment can be polymerised as a random mixture of two structural forms, the 1,4 and 1,2 isomers, by addition of an inert polar material to the polymerisation solvent. Ethers and amines have been suggested for this purpose. Upon hydrogenation, these isomers give a copolymer of ethylene and butylene. The S–EB–S block copolymers produced in this way have excellent resistance to degradation. Similarly, S–I–S block copolymers can be hydrogenated to give the more stable S–EP–S equivalents.

Both traditional and living anionic polymerisation remain an important technique for the preparation of well-defined triblock copolymers, such as poly(styrene-b-butadiene-b-styrene) and poly(styrene-b-isoprene-b-styrene). They are extended also to copolymers containing polysiloxane blocks or to poly(α-methylstyrene-b-propylene sulphide-b-α-methylstyrene).

The use of living cationic polymerisation in the preparation of TPEs was reviewed by Kennedy [30, 31] in relation to graft and block copolymers, but the application of cationic polymerisation to TPEs began before the arrival of the living techniques, based on a three-step progression: (i) controlled initiation, (ii) reversible termination (quasi-living systems) and (iii) controlled transfer [32]. Cationic polymerisation is more complex than the anionic synthesis, and it has been used to produce block copolymers with polyisobutylene mid-segments, or poly(styrene-b-isobutylene-b-styrene) (S-I-B-S). Polyisobutylene is the only mid segment that can be produced by this method while there are many aromatic polymers that can form the end segments. Anionic and cationic polymerisations are often associated. Feldthusen et al. prepared copolymers containing linear and star-shaped blocks: a living polyisobutylene chain was prepared by cationic polymerisation, its ends were converted into 2,2-diphenylvinyl groups, then used as initiators of the *tert*-butyl methacrylate anionic polymerisation [33].

The discovery of the controlled radical polymerisation (CRP) offered additional possibilities to the chemistry of TPEs [34–36]. CRP was used in both graft and block copolymer preparation and extensively reviewed by Matyjaszewski [37].

Polyolefin TPEs are produced using metallocene catalysts, typically based on cyclopentadienyl groups linked to the halide of a transition metal (i.e., Ti, Zr, Hf). Under the right conditions, they polymerise mixtures of ethylene and α-olefin monomers (usually 1-octene) into polymers with long, repeating polyethylene segments that form the hard phase in the polymer. There are also copolymer segments with pendant groups, usually arranged atactically. Because of their random atactic structures, these segments cannot crystallise, so they are amorphous materials with low T_g and rubber-like at room temperature, thus forming the soft phase in the polymer.

The development of graft copolymers is by far less important than that of block copolymers. Grafting is mainly used to modify the properties of a block copolymer. It may proceed through two different pathways: direct reaction of the backbone with a monofunctional oligomer (grafting onto) or polymerisation of a monomer initiated by an active group of the polymer (grafting from).

5.5 Property–structure relationships

Since most TPEs are phase-separated systems, they show many of the characteristics of the individual polymers that constitute each phase [1, 2]. For example, each phase has its own T_g, or the melting point (T_m), if it is semi-crystalline. These, in turn, determine the temperatures at which a particular TPE goes through transitions in its physical properties.

When the modulus of a TPE is measured over a range of temperatures, there are three distinct regions, as shown in Fig. 5.2. At very low temperatures, both phases are hard and so the material is stiff and brittle. At a somewhat higher temperature, the elastomer phase becomes soft and the TPE now resembles a conventional vulcanisate. As the temperature is further increased, the modulus stays relatively constant (the region is often described as rubbery plateau) until finally the hard phase softens. At this point, the TPE becomes fluid.

TPEs have two service temperatures. The lower service temperature depends on the T_g of the elastomer phase, while the upper service temperature depends on the T_g or T_m of the hard phase. Values of T_g and T_m for the various phases in some commercially important TPEs are given in Tab. 5.3.

Some of the parameters that can be varied in order to change the properties of TPEs include:

- *Molecular Weight.* Compared with homopolymers of similar molecular weight, styrenic block copolymers have very high melt viscosities that increase with the molecular weight. These effects are attributed to the persistence of the two-phase domain structure in the melt and the extra energy required to disrupt this structure during flow. If the styrene content is held constant, the total molecular weight has little or no effect on the modulus of the material at

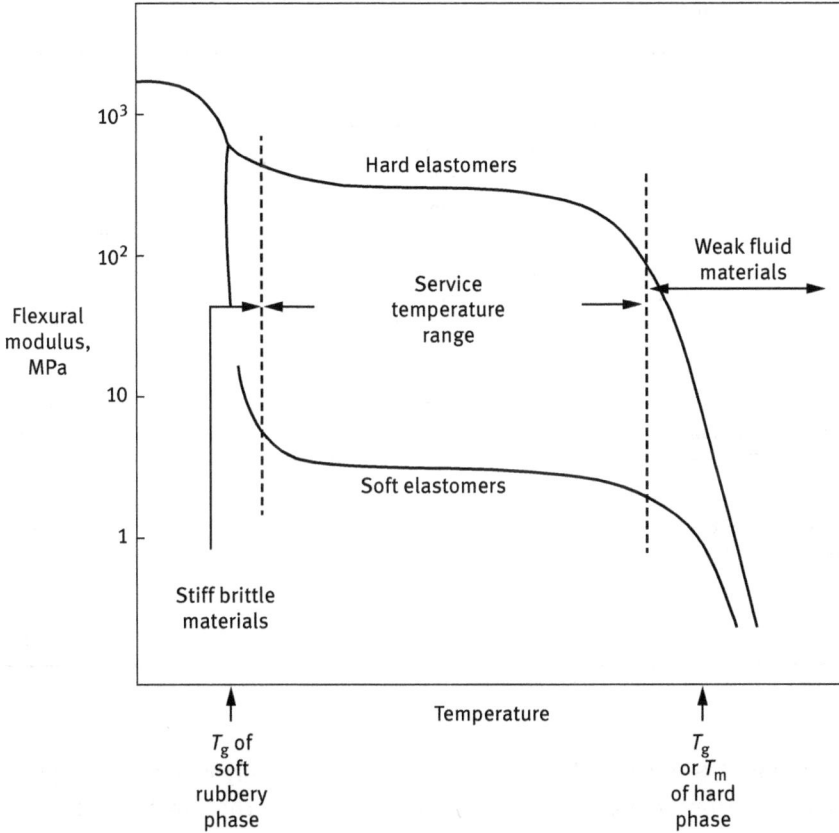

Fig. 5.2: Modulus behaviour vs temperature for a typical TPE.

ambient temperature. This is attributed to the modulus of the elastomer phase, that is inversely proportional to the molecular weight between the entanglements in the elastomer chains, since this value is not affected by the total molecular weight.

– *Proportion of hard phase.* As the ratio of the hard to soft phase is increased, polymers in turn become harder. Indeed, they change from very weak, soft, rubber-like materials to strong elastomers, then to leathery materials and finally to hard flexible thermoplastics (i.e., high impact polystyrenes).

– *Elastomer Phase.* The choice of the elastomeric segment has a remarkable effect on the properties of block copolymers. In the styrenic block copolymers, four elastomers are commercially important: polybutadiene, polyisoprene, poly(ethylene-co-butylene) and poly(ethylene-co-propylene). Polyisobutylene has also been extensively investigated but the products have not been commercially produced. The corresponding styrenic triblock copolymers are referred to as S–B–S,

Tab. 5.3: Glass transition and crystal melting temperatures of some TPEs.

TPE	Soft rubbery phase T_g (°C)	Hard phase T_g or T_m (°C)
Polystyrene/elastomer block copolymers		
SBS	−90	95 (T_g)
SIS	−60	95 (T_g)
SEBS and SEPS	−60	95 (T_g) and 165 (T_m)
Multiblock copolymers		
Polyurethane/elastomer	−40−−60	190 (T_m)
Polyester/elastomer	−40	185–220 (T_m)
Polyamide/elastomer	−40−−60	220–275 (T_m)
Polyethylene/poly(α-olefin)	−50	70 (T_m)
Polyetherimide/polysiloxane	−60	225 (T_g)
Hard polymer/elastomer blends		
Polypropylene/EPDM or EPR	−50	165 (T_m)
Polypropylene/butyl rubber	−60	165 (T_m)
Polypropylene/natural rubber	−60	165 (T_m)
Polypropylene/nitrile rubber	−40	165 (T_m)
PVC/nitrile rubber/DOP	−30	80 (T_m)

S–I–S, S–EB–S, S–EP–S and S–iB–S, respectively. Both polybutadiene and polyisoprene have one double bond per monomeric unit. The double bond is a source of instability, thus limiting the thermal and oxidative stability of the S–I–S and S–B–S block copolymers. In contrast, poly(ethylene-co-butylene) and poly(ethylene-co-propylene) are completely saturated, and so S–EB–S and S–EP–S block copolymers are much more stable.

Another important aspect is the modulus of the materials. It is postulated that the modulus of styrenic block copolymers is inversely proportional to the molecular weight between chain entanglements (Mc), as well as to the effects of the polystyrene domains acting as reinforcing filler particles. Values of Mc are as follows: polyisobutylene, 8900; natural rubber, 6100; polybutadiene, 1700 and poly(ethylene-co-propylene), 1660. The Mc for poly(ethylene-co-butylene) is similar to that of poly(ethylene-co-propylene). Because of these differences in Mc, S–iB–S block copolymers are the softest of all, S–I–S block copolymers are softer than the S–B–S analogues and the S–EB–S and S–EP–S analogues are the hardest.

Poly(ethylene-co-butylene) and poly (ethylene-co-propylene) are nonpolar. The corresponding block copolymers can thus be compounded with hydrocarbon-based extending oils, but do not have much oil resistance. Conversely, block copolymers with polar polyester or polyether elastomer segments have little affinity for such hydrocarbon oils and so have better oil resistance.

Among polyurethane, polyester and polyamide TPEs, those with polyether-based elastomer segments show better hydrolytic stability and low temperature flexibility, whereas the polyester-based analogues are tougher and have the best oil resistance. Polycaprolactones and aliphatic polycarbonates, two special types of polyesters, are used to produce premium-grade polyurethanes.

- *Hard phase.* The choice of the hard phase determines the upper service temperature and also influences the solvent resistance. In styrenic block copolymers, those based on poly(α-methylstyrene) have higher upper service temperature and tensile strength than the analogues based on polystyrene. Replacing the polystyrene end segments in S–EB–S by polyethylene (giving E–EB–E block copolymer) improves the solvent resistance.

In polyurethane/elastomer, polyester/elastomer and polyamide block copolymers, the crystalline end segments, together with the polar centre segments, impart good oil resistance and high upper service temperatures. In the polyolefin block copolymers, polyethylene is the hard phase. It has a relatively low T_m (about 70 °C) and so these TPEs should have a low upper service temperature. The hard segment in the polyetherimide/polysiloxane block copolymers has a very high T_g (about 225 °C) and so these TPEs have a very high upper service temperature.

Polypropylene is used as the hard phase in many hard polymer/elastomer combinations. It is low in cost and density and its T_m is quite high (about 165 °C). This crystalline polypropylene phase imparts resistance to solvents and oils, as well as relatively high upper service temperatures.

5.6 Dynamic vulcanisation

Dynamic vulcanisation is a widely used method to prepare TPEs comprising partially or fully cross-linked elastomer particles in a melt-processable thermoplastic matrix [1, 2]. Dynamic vulcanisation is thus defined as the process of cross-linking rubber during its intimate melt mixing with a non-vulcanising thermoplastic polymer. This occurs simultaneously in an internal mixer or in a twin-screw extruder. The resulting materials, called thermoplastic vulcanisates (TPV), are very elastomeric in their performance characteristics.

The difference between the TPOs and TPVs is that both the elastomeric and polyolefin phases in TPOs are co-continuous phases while in the TPVs the polyolefin phase is continuous and surrounds the cross-linked and discontinuous elastomeric phase. A two-dimensional representation of TPOs and TPVs is shown in Fig. 5.3. The TPV morphology has many common characteristics with the TPE polyolefins,

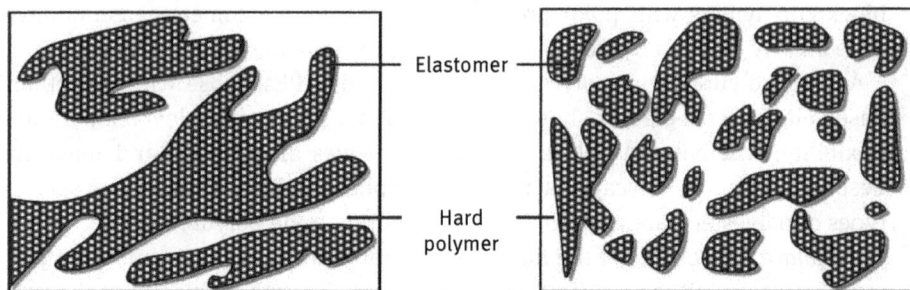

Fig. 5.3: Development of morphology in a thermoplastic vulcanisates from a co-continuous phase on the left (TPO) to a dispersed phase on the right (TPV).

involving a highly vulcanised elastomeric phase uniformly distributed in a melt-processable matrix.

There are three basic requirements in order to obtain TPVs with good balance of properties: i) the wetting surface tensions of both the elastomeric and thermoplastic components, ii) the crystallinity fraction of the thermoplastic phase and iii) the critical entanglement spacing of the elastomer macromolecules. The best combinations of elastomer and thermoplastic are those in which the surface energies of the two components are matched, the entanglement molecular weight of the elastomer is low and the thermoplastic is at least 15% crystalline. As example, a decrease in both tensile strength and elongation at break has been found with an increase in wetting surface tension and critical entanglement spacing and a decrease in the crystallinity fraction.

The most commonly used combinations are based on dynamically vulcanised EPDM and polyolefins. However, it is possible to make TPVs with a variety of thermoplastics and elastomers. Other blends include butyl and halobutyl rubbers, polyacrylate rubber and polyolefins and butadiene-acrylonitrile rubber and PVC [38].

The high incompatibility between the elastomer and plastic phases may be an important obstacle in the preparation of a dynamically vulcanised material, since the properties of the latter depend on the quality of the dispersion. So, a polymeric compatibiliser, which often requires grafting and coupling processes, must be added [24, 39]. Compatibilisers enhance the inter-phase reaction between the elastomeric and plastic phases, which result in the mutual wetting of the two phases due to which the blends became more compatible.

Dynamic vulcanisation allows to produce TPEs that have properties as good or even better, in some cases, than those of block copolymers. The ratio between the elastomeric and thermoplastic components in the systems is crucial for the main final properties (i.e., tensile strength, elongation at break, tear strength and resilience and so on). If the elastomeric particles are small enough and if they are fully vulcanised, the properties of the blends are greatly improved. These improvements include enhanced ultimate mechanical properties, reduced permanent set, improved fatigue resistance, greater impact resistance, improved high temperature utility, greater melt strength, more reliable thermoplastic fabricability and so on.

TPVs have taken the place of thermoset rubber in many applications since they entered the market 20 years ago. At the beginning, TPVs were used to substitute existing applications of elastomers, now they are also opening new fields of application due to their processing potential. TPVs play an important role in automotive industry as vacuum tubing, body plugs, seals, air conditioning hose cover, emission tubing and fuel line hose cover. They are widely used in electrical applications as wire and cable insulation and jacketing, connectors and terminal ends. Furthermore, TPVs can also be used in mechanical rubber goods applications such as convoluted bellows, mount, bumpers, housings and oil well injection lines, and for window and door impermeability, instead of thermoset rubbers like EPDM.

5.7 Styrenic block copolymers

Styrenic block copolymers are the largest volume and of the lowest priced category of TPEs [1, 2]. Generally, they offer a much wider range of properties than conventional cross-linked rubbers because the composition in the block nature and their ratio can vary to suit the customer needs. As already mentioned in Section 5.3, they are based on simple molecules (A–B–A type) that consist of at least three blocks, namely two hard polystyrene end blocks and one soft elastomeric midblock. The midblock is typically a polydiene, either polybutadiene or polyisoprene, resulting in the well-known family of styrene–butadiene–styrene (S–B–S) and styrene–isoprene–styrene (S–I–S).

Styrenic block copolymers are technologically compatible with a surprisingly wide range of materials and can be blended to give useful products. Blends of S–B–S with polystyrene, polyethylene or polypropylene show improved impact and tear resistance. Both S–B–S and S–EB–S can be blended with poly(phenylene oxide) to improve the impact resistance. S–EB–S can also be blended with the less polar engineering thermoplastics such as polycarbonates. Another development is the use of functionalised S–EB–S block copolymers as impact modifiers for more polar thermoplastics such as polyesters and polyamides. The functionality is given by maleic acid/anhydride groups grafted to the S–EB–S polymer chain.

Commercial products have hardness from 5 on the Shore A scale (which is extremely soft) to 45 on the Shore D scale (almost leathery). Specific gravities usually range from 0.9 to 1.20 g/cm^3; some products intended for soundproofing have specific gravities as high as 1.95 g/cm^3. Processing is relatively easy. In general, products based on S–B–S are processed under conditions appropriate for polystyrene, whereas products based on S–EB–S are processed under conditions appropriate for polyethylene. Pre-drying is usually not needed and scrap is recycled.

In all their commercial applications, the styrenic block copolymers are never used as pure materials. Instead, they are compounded with oils, fillers, resins and so on, to give the materials designed for the specific end uses. The effects of compounding ingredients on the properties of styrenic block copolymers are summarized in Tab. 5.4. Oils with high aromatic content should be avoided because they plasticise the polystyrene domains. Polystyrene is often used as an ingredient in S–B–S-based compounds; it makes the products harder and improves their processability. In S–EB–S-based compounds, crystalline polyolefins such as polypropylene are preferred.

Tab. 5.4: Compounding styrenic block copolymers.

Property	Component				
	Oils	*PS*	*PE*	*PP*	*Fillers*
Hardness	Decreases	Increases	Increases	Increases	Small increase
Processability	Improves	Improves	Improves	Improves, especially in SEBS	Improves
Effect on oil resistance	None	None	Improves	Improves	None
Cost	Decreases	Decreases	Decreases	Decreases	Decreases

Special grades of styrenic block copolymers are useful modifiers for sheet moulding compounds based on thermoset polyesters, since they improve the surface appearance, impact resistance and hot strength. Other uses for which special compounds have been developed include materials intended for food contact, wire insulation and pharmaceutical applications.

5.8 Polyolefin blends

TPEs based on polyolefins (TPO), like polyethylene, are considered to be an important family of engineering materials [29, 40]. They consist of a polyolefin semi-crystalline thermoplastic, providing the strength, and an amorphous elastomeric component, providing the flexibility. TPO ingredients generally include ethylene/propylene random copolymer, isotactic polypropylene and the addition of other fillers and additives. The combination of rubbery properties along with the thermoplastic nature of polyethylene/elastomer blends make them as very important materials in industrial applications, thanks to their useful properties.

TPOs containing high amounts of elastomers are quite rubbery with high elongation at break values, while TPOs containing high amounts of polyolefin undergo a small recovery after stress. They are available in the hardness range from 60 Shore A to 70 Shore D. Hard TPO compounds are often used in automotive. Soft TPO compounds can be extruded into a sheet and thermoformed for large automotive part such as interior skin. TPO chemical composition results in good resistance to many solvents and ozone. They can be stabilised for good outdoor ageing. TPOs are excellent electrical insulating materials as well.

5.9 Thermoplastic elastomers based on polyamide

Thermoplastic polyamide elastomers (COPAs) belong to the group of segmented block copolymer showing a structure with repeating hard and soft segments [1, 2]. The hard segments are polyamides that serve as virtual cross-links reducing the chain slippage and viscous flow of the copolymer, whilst the soft segments are either polyethers or polyesters that contribute to the flexibility and extensibility of elastomers. Both segments are connected by amide linkages.

The most important members of COPAs are polyester amides (PEAs), polyether ester amides (PEEAs), polycarbonate ester amides (PCEAs) and polyether-block-amides (PE-b-As). The hard segment controls the degree of crystallinity, the crystalline T_m and the mechanical strength while the soft segment determines the thermal oxidative and hydrolytic stability, chemical resistance and low temperature flexibility.

The properties of COPAs may vary according to some factors as: proportion of the hard and soft segments in the copolymer, chemical composition, molecular weight distribution, method of preparation and thermal history (affecting the degree of phase separation and domain formation).

Most of COPAs exhibit higher resistance to elevated temperature than any other commercial TPEs. They are also higher resistant to long-term dry heat ageing without adding any heat stabilisers. COPAs have excellent abrasion resistance, which is comparable to that of TPUs. The hardness is in the range from 80 Shore A to 70 Shore D by varying the content of hard and soft segments. The good insulation properties of COPAs make them suitable for low voltage applications and for jacketing. Other application areas for these materials include conveyor and drive belt, footwear such as ski boots and sport shoes, automotive applications, electronics, hot melt adhesives, powder coatings for metals and impact modifiers for engineering thermoplastics.

Polyester amides constitute a peculiar of biodegradable family, due to the presence of both ester and amide groups that guaranties degradability. These biodegradable polymers are receiving great attention and are currently being developed for a great number of biomedical applications such as controlled drug delivery systems, hydrogels, tissue engineering and other uses.

5.10 Thermoplastic polyether ester elastomers

Polyether ester elastomers or copolyesters (COPEs) consist of a sequence of hard and soft segments [1, 2]. The high melting blocks (hard segments) are formed by the crystalline polyester segments that are capable of crystallisation, whereas the rubbery soft segments are formed by the amorphous polyether segments with a relatively low T_g. At useful service temperatures, the polyester blocks form crystalline domains embedded in the rubbery polyether continuous phase. These crystalline domains act as the physical cross-links. At elevated temperatures, crystallites break down to yield a polymer melt, thus facilitating the thermoplastic processing.

COPEs are considered engineering TPEs because of their unusual combination of strength, elasticity and dynamic properties. They have a wide useful temperature range between the T_g (around –50 °C) and T_m (around 200 °C). These materials are elastic but their recoverable elasticity is limited to low strains. They have excellent dynamic performance and show resistance to creep. Because of their high modulus and stiffness, they have been used to replace some conventional rubbers, PVC and other plastics in many applications. COPEs are resistant to oils, aliphatics and aromatic hydrocarbons, alcohols, ketones, esters and hydraulic fluids. Thus, uses of COPEs are reported in fuel tanks, quiet running gear wheels, hydraulic hoses, tubing, seals, gaskets, flexible couplings, wire and cable jacketing. COPEs can also be used in electrical applications for voltages 600 V and less.

5.11 Thermoplastic polyurethane

Thermoplastic polyurethane (TPU) is a unique category of plastic, synthesised by a polyaddition reaction between a diisocyanate and one or more diols, extremely popular across a range of markets and applications [1, 2]. First developed in 1937, this versatile polymer is soft and processable when heated, hard when cooled and capable of being reprocessed multiple times without losing its structural integrity.

Inherently flexible, TPU can be extruded or injection moulded on conventional thermoplastic manufacturing equipment to create solid components typically for footwear, cable and wire, hose and tube, film and sheet or other industry products. It can also be compounded to create robust plastic items or processed using organic solvents to form laminated textiles, protective coatings or functional adhesives.

There are three main chemical classes of TPU: polyester, polyether and a smaller class known as polycaprolactone. Polyester TPUs are compatible with PVC and other polar plastics. Offering value in the form of enhanced properties, they are unaffected by oils and chemicals, provide excellent abrasion resistance, give a good balance of physical properties and are perfect for use in polyblends. Polyether TPUs are slightly lower in specific gravity than polyester and polycaprolactone grades. They offer low temperature flexibility and good abrasion and tear resilience.

They are also durable against microbial attack and provide excellent hydrolysis resistance, making them suitable for applications where water is used. Polycaprolactone TPUs have the inherent toughness and resistance of polyester-based TPUs combined with low-temperature performance and a relatively high resistance to hydrolysis. They are an ideal raw material for hydraulic and pneumatic seals.

Structurally, a TPU is a multi-phase block copolymer obtained by reaction between a polyol (soft segment) and a diisocyanate (hard segment). Generally, two diols are required: a chain extender or short chain diol (i.e., 1,4-butanediol) and an elastomeric hydroxyl terminated polymer (polyoil or long-chain diol, also called chain extender), such as polyethers (i.e., polyoxyethylene, polyoxybutylene), polyesters (i.e., polyethylene succinate, polybutylene succinate, polyethylene adipate, polybutylene), hydroxyl terminated polybutadiene and hydroxyl terminated poly (butadiene-*co*-acrylonitrile). The most used diisocyanates are methane 4,4′-diphenyl diisocyanate (MDI), 2,4- and 2,6-toluene diisocyanate (TDI) and 1,6-hexane diisocyanate (HDI).

Looking at the isocyanate nature, TPUs can also be subdivided into aromatic and aliphatic varieties:
- Aromatic TPUs: Based on isocyanates like MDI and TDI, they are workhorse products and can be used in applications that require flexibility, strength and toughness.
- Aliphatic TPUs: Based on isocyanates like HDI, they are light stable and offer excellent optical clarity. They are commonly employed in automotive interior and exterior applications and as laminating films to bond glass and polycarbonate together in the glazing industry. They are also used in projects where attributes like optical clarity, adhesion and surface protection are required.

Polyurethane block copolymers are produced from prepolymers by polycondensation following the one-step or two-step method [29]. Sequences are found to be more regular in the polymer obtained with the two-step method. The structural regularity leads to a better packing of hard segments where physical cross-linking points are easier to form. Hence, the two-step method gives a product with better mechanical properties than the one-step method does. Again, the solubility of these two products is different. The polyurethanes obtained from one-step method are soluble in some of the common solvents, but the polyurethanes from the two-step process could not be dissolved in any common solvents.

In the two-step method, a relatively high molecular weight polyester or polyether with terminal hydroxy groups (a polyglycol) first reacts with an excess of diisocyanate:

$$HOR_1OH + 2OCNR_2NCO \rightarrow OCNR_2NHCOOR_1OOCNHR_2NCO$$

This reaction continues to give a prepolymer in which the segments derived from the polyglycol and diisocyanate alternate, and which is terminated by isocyanate groups. This prepolymer, designated $OCNR_3NCO$, cannot crystallise because it has an irregular structure. It, in turn, reacts with a low molecular weight glycol, such as 1,4-butanediol, and with more diisocyanate forming the final polymer:

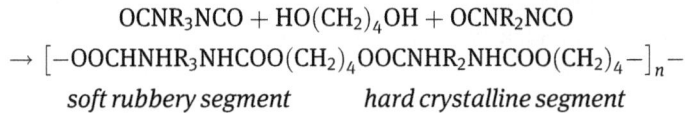

$$OCNR_3NCO + HO(CH_2)_4OH + OCNR_2NCO$$
$$\rightarrow \left[-OOCHNHR_3NHCOO(CH_2)_4OOCNHR_2NHCOO(CH_2)_4-\right]_n-$$

soft rubbery segment 　　　*hard crystalline segment*

Fig. 5.4 shows the schematic structure of TPUs. The soft block is responsible for the flexibility and elastomeric character. The hard block imparts its properties of toughness and physical performance to the material. The hard and soft chain segments phase separate with the hard segments as a dispersed minor phase, since the soft segments form the continuous phase.

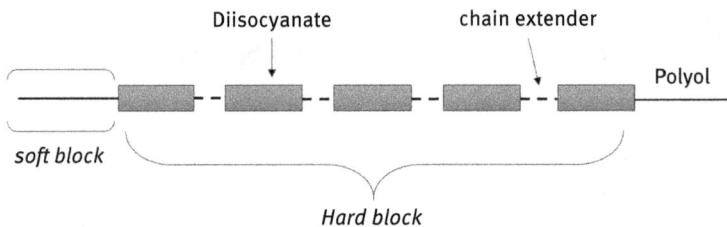

Fig. 5.4: Schematic structure of TPUs.

References

[1] Holden, G., Legge, NR., Quirk, RP., Schroeder, HE. Thermoplastic elastomers. 2nd ed., Hanser & Hanser/ Gardner, Munich, Germany, 1996.
[2] Holden, G. Understanding thermoplastic elastomers. Hanser & Hanser/ Gardner, Munich, Germany, 2000.
[3] Baumann, m. In Proceedings of thermoplastic elastomers topical conference. Houston, Tx, USA, 2001, p. 99.
[4] Lieser, TH., Marcura, K. Künstliche organische Hochpolymere I. Über die Reaktionsweise von Acyl-diisocyanaten mit polyfunktionellen Amino-und Hydroxylverbindungen. Annalen 1941, 548, 226.
[5] Bayer, O. Bemerkungen zu der Abhandlung von Th. Lieser und Karl Macura: Künstliche organische Hochpolymere. I. Annalen 1941, 549, 286.
[6] Bayer, O., Rinke, H., Siefken, L., Orthner, L., Schild, H. German patent DE 728, 981, 1937.
[7] Christ, RE., Hanford, WE. US patent 2,333,639, 1943.
[8] Coffey, DH., Cook, JG., Lake, WH. British patent GB 574 134, 1945.
[9] Müller, E., Petersen, S., Bayer, O. German patent DE 76 584, 1943.
[10] Petersen, S., Müller, E., Bayer, O. German patent DE 77 229, 1944.
[11] Snyder, MD. US patent 2,623,031, 1952.

[12] Szwarc, M., Levy, M., Milkovich, R. Polymerization initiated by electron transfer to monomer. A new method of formation of block polymers. J. Am. Chem. Soc. 1956, 78, 2656.

[13] Szwarc, M. 'Living' polymers. Nature 1956, 178, 1168.

[14] Bailey, JT., Bishop, ET., Hendricks, WR., Holden, G., Legge, NR. Thermoplastic elastomers-physical properties and applications. Rubber Age. 1966, 98, 69.

[15] Holden, G., Bishop, ET., Legge, NR. Thermoplastic elastomers. J. Polym. Sci. Part C Polym. Symp. Ed. 1969, 26, 37.

[16] Schollenberger CS., Scott, H., Moore, GR. Polyurethan VC, a virtually crosslinked elastomer. Rubber World. 1958, 137, 549.

[17] Kontos, EG., Easterbrook, EK., Gilbert, RD. Block copolymers of α-olefins prepared from macromolecules with long chain lifetimes. J. Polym. Sci. 1962, 61, 69.

[18] Tobolsky, AV. Trends in rubber research. Rubber world. 1959, 139, 857.

[19] Hartman, PF., Eddy, CL., Koo GP. A report on new elastomeric thermoplastics. SPE Journal, 1970, 26, 62.

[20] Rodriguez, F., Winding, CC. Mechanical degradation of dilute polyisobutylene solutions. Ind. Eng. Chem. 1959, 51, 1281.

[21] Chen, AT., Nelb, RG., Onder, K. New high-temperature thermoplastic elastomers. Rubber Chem. Technol. 1986, 59, 615.

[22] Deleens, G., Foy, P., Maréchal, E. Synthese et caracterisation de copolycondensats sequences poly(amide-seq-ether)-I. Synthese et etude de divers oligomeres ω,ω' difonctionnels du poly(amide-11). Eur. Polym. J. 1977, 13, 337.

[23] Nishimura, AA., Komagata, H. Elastomers based on polyester. J. Macromol. Sci–Chem. 1967, A1, 617.

[24] Roy Choudhury, N., Bhowwick, AK. Compatibilization of natural rubber–polyolefin thermoplastic elastomeric blends by phase modification. J. Appl. Poly. Sci. 1989, 38, 1091.

[25] Michaeli, M., Kautsch, W. Morphology and mechanical properties of natural rubber/polyethylene blends. Gummi. Kunstst. 1991, 44, 827.

[26] Rader, CP. In Elastomer technology – Compounding and testing for performance, J. Dick Ed., Hanser Publisher, Munich, Germany, 2001, 10, p. 264.

[27] France, C, Hendra, PJ, Maddams, WF, Willis, HA. A study of linear low-density polyethylenes: branch content, branch distribution and crystallinity. Polymer. 1987, 28, 710.

[28] Kuriakosc, B., De, SK. Dynamic mechanical properties of thermoplastic elastomers from polypropylene–natural rubber blend. J. Appl. Polym. Sci. 1986, 32, 5509.

[29] Chu, SG., Class, J. The viscoelastic properties of rubber–resin blends. I. The effect of resin structure. J. Appl. Polym. Sci. 1985, 30, 805.

[30] Kennedy, JP., Maréchal, E. In Carbocationic polymerization, Wiley-Interscience, New York, USA, 1992, 8, p. 410.

[31] Kennedy, JP. Living cationic polymerization of olefins. How did the discovery come about? J. Polym. Sci. Part A Polym. Chem. 1999, 37, 2285.

[32] Abdullah, I., Ahmad, S., Sulaiman, CS. Blending of natural rubber with linear low-density polyethylene. J. Appl. Polym. Sci. 1995, 58, 1125.

[33] Feldthusen, J., Bela, I., Mueller, AHE. Synthesis of linear and star-shaped block copolymers of isobutylene and methacrylates by combination of living cationic and anionic polymerizations. Macromolecules. 1998, 31, 578.

[34] Matyjaszewski, K., Gaynor, S., Greszta, D., Mardare, D., Shigemoto, T. Synthesis of well defined polymers by controlled radical polymerization. Macromol. Symp. 1995, 98, 73.

[35] Matyjaszewski, K., Gaynor, S., Wang, JS. Controlled radical polymerizations: the use of alkyl iodides in degenerative transfer. Macromolecules. 1995, 28, 2093.

[36] Ameduri, B., Boutevin, B., Gramain, P. Synthesis of block copolymers by radical polymerization and telomerization. Adv. Polym. Sci. 1997, 127, 87.

[37] Matyjaszewski, K. Environmental aspects of controlled radical polymerization. Macromol. Symp. 2000, 152, 29.

[38] Soares, BG., Santos, DM., Sirqueira, AS. A novel thermoplastic elastomer based on dynamically vulcanized polypropylene/acrylic rubber blends. Express Polym. Lett. 2008, 2, 602.

[39] Dahlan, HH., Zaman, MDK., Ibrahim, A. Liquid natural rubber (LNR) as a compatibilizer in NR/LLDPE blends. J. Appl. Polym. Sci. 2000, 78, 1776.

[40] Ha, CS., Ihm, DJ., Kim, SC. Structure and properties of dynamically cured EPDM/PP blends. J. Appl. Polym. Sci. 1986, 32, 6281.

6 Additives and fillers

6.1 Introduction

In its original state, an elastomer is generally not very strong, does not maintain its shape after a large deformation, can be very sticky with a consistency of a chewing gum, has limited resistance to solvents, is prone to the attack by oxygen, ozone and so on [1, 2]. Useful products cannot be made without any modifying raw rubbers since they are too weak to fulfil the practical requirements due to the lack of hardness, strength properties, wear resistance and so on. It is, therefore, necessary to mix the elastomer with certain additives to optimize its properties to meet a given application or to set up the performance characteristics. Additives must be environmentally safe, processable in manufacturing facilities and cost effective.

As discussed in Chapter 2, the process of mixing various ingredients with elastomers, each of them having a specific function in processing, vulcanisation or end uses, is known as compounding. The formulation thus developed is called as a *recipe*. All the ingredients included in every formulation are normally given in amounts based on a total of 100 parts of rubber (phr).

Additives are added as per well-defined steps during the production of the compound, depending on their function in the raw rubber: they can be adde d in the mastication step, in the first masterbatch or during the last mixing step. In any case, the first goal of any compounding run is to achieve the best possible dispersion and distribution in the elastomeric matrix. A good distribution can be achieved if the additive softens below the processing temperature (mill mixer: max. 70 °C and internal mixer: max. 135 °C) and melts down at temperatures below the processing temperature. For example, sulphur or metal oxides do not melt during mixing, so sulphur and metal oxide pre-dispersions in a low-melting matrix excellently distribute the material during mixing. Pre-dispersions may help to carry the non-melting product into the compound, preventing the re-agglomeration of additive, thus ensuring an excellent dispersion in critical compounds.

The major classes of additives have already been discussed in Chapter 2 (Section 2.3), and those related to the vulcanising system, including curatives, accelerators, activators and retarders, have been described in Chapter 3. In this chapter, some highlights especially on fillers, anti-degradants and lubricants will be discussed.

6.2 Fillers

The use of fillers in rubber is almost as old as the use of rubber itself. In the past, when the polymer reinforcement was not the main concern, different types of clay

https://doi.org/10.1515/9783110640328-006

minerals were used as fillers to reduce the cost of the host polymer and provide certain properties useful for rubber compounding [1–4]. In this sense, filler can be considered a diluent used primarily to lower the volume cost, even though with use, all fillers can modify certain physical properties of the compound. Generally, lower cost is achieved at the expense of other desirable properties. Later, it has been observed that the incorporation of particulate fillers, such as carbon black (CB), could significantly increase the strength of vulcanised rubbers.

The selection of fillers for a given formulation is based on the property requirements of the end-products. Chemical composition and its effects on the compound physical properties typically result in classifying fillers into three broad categories based on their reinforcement action [5]:

1. *Non-reinforcing*: They show small or no physical interactions with the elastomeric phase; thus, they do not improve the mechanical properties and work only as diluents [6]. Examples include clay (kaolin) and calcium carbonate. Clay, which has been widely used as cheap filler in the rubber industry, has poor reinforcing capability because of its large particle size and low surface activity.
2. *Semi-reinforcing or extending fillers*: They are able to moderately improve the tensile strength and the tear strength, but do not improve the abrasion resistance.
3. *Reinforcing fillers*: They create physical and/or chemical interactions with the elastomeric matrix, thus affecting vulcanisate performances and improving stiffness, modulus and failure properties (tensile strength, tear resistance and abrasion resistance). CB and precipitated silica belong to this family.

Generally, particulate fillers used in rubber industry can also be classified as *black* and *non-black* (sometime called *white*) depending on their origin. The former are mostly produced from petroleum feed stock (CB), whereas the latter from mineral sources (i.e., silica, silicates, clays, calcium carbonate and other mineral fillers, used extensively where a high degree of reinforcement is not essential) [6].

CB is the most common and effective reinforcing filler, thanks to its unique ability to significantly enhance the properties of nearly any elastomer system. During vulcanisation, CB enters into the chemical reactions with sulphur and accelerators participating in the formation of the vulcanised network, thus influencing the rate of vulcanisation and the degree of cross-linking. The cross-link density of a black-reinforced vulcanisation system increases about 25% in comparison with the corresponding unfilled one. As a consequence, the stress–strain properties are enhanced, whereas the rubber swelling in organic solvents is generally reduced.

About 5 million metric tons of CB is globally consumed each year, while only 250,000 tons of the different silica grades, including the highly dispersible silica, are used annually. However, because of its polluting nature, the ubiquitous black colour of the compounded rubber and its dependence on petroleum feedstock for the synthesis cause researchers to look out for other *white* reinforcing materials. Precipitated silica is the best non-black reinforcing filler so far developed, with a particle size as fine as that of CB and an extremely good reactive surface. The proper choice of the vulcanising systems allows to obtain appropriate scorch and cure times in silica and silicate-filled compound. A distinct advantage imparted by silica to many rubbers is the increased resistance to air ageing at elevated temperatures.

The effect of reinforcing agents on the properties of the rubber–filler mixture is of great practical significance and, for the manufacture of some items, the selection of the most suitable reinforcing agents may be a more important factor than their action on the vulcanisate's properties. In general, the finer reinforcing agents require more energy for their dispersion into rubber and the plasticity of the resultant mix is lower. Therefore, these compounds are more difficult to process in the operations following mixing. On the other hand, high-structure fillers incorporate more slowly than low-structure fillers, but once incorporated, the former disperses more easily and rapidly than their low-structure counterpart.

6.2.1 Filler characteristics

The effect of fillers on the reinforcement of rubber vulcanisates depends on some factors strictly related to the nature of the filler, described as follows [5–8].

Particle size

The reinforcing effect of any filler is strictly linked to the particle size. Since the particle size is directly related to the reciprocal of the surface area per gram of filler, an increase in the surface area that is in contact with the rubber phase leads to an increase in reinforcement. Reducing the particle size simply results in a greater influence on polymer–fillers interactions.

When the size of the filler particle greatly exceeds the polymer interchain distance, an area of localised stress, which can contribute to the elastomer chain breaking on flexure or stretching, is introduced in the material. Fillers with particle size greater than 10,000 nm (10 μm) are therefore generally avoided because they can reduce performances rather than extend or reinforce (Fig. 6.1). The presence of large particles or agglomerates in the rubber not only reduces the contact between filler and rubber matrix but also works as failure initiation sites, which would lead to the premature failure of materials. Fillers with particle size between 1,000 and 10,000 nm (1–10 μm) are used primarily as diluents and usually have no significant effect, positive or negative, on properties of rubber. Semi-reinforcing fillers range from 100 to 1,000 nm (0.1–1 μm). The truly reinforcing fillers, which range from 10 to 100 nm (0.01–0.1 μm), significantly improve properties of rubber.

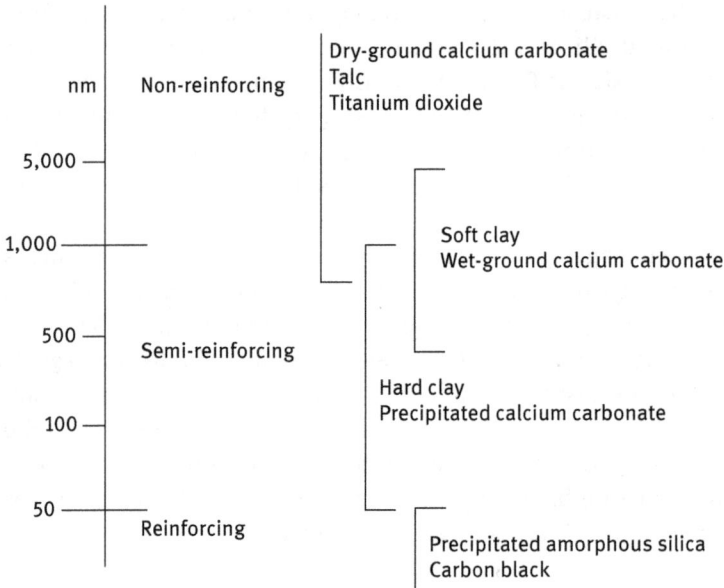

Fig. 6.1: Classification of fillers based on their reinforcing effect and the particle size.

The increase of modulus and tensile strength is strongly dependent on the particle size of the filler: smaller particle size fillers impart greater reinforcement to the rubber compound than the coarse ones. However, different fillers of the same particle size may not impart the same reinforcement, as in the case of CB and silica.

In addition to the average particle size, the particle-size distribution also has a significant effect on reinforcement. Fillers with a broad particle-size distribution show better packing in the rubber matrix, which results in a lower viscosity than that provided by an equal volume of fillers with a narrow particle-size distribution.

Particle geometry

Shape and geometry of the filler particle are important factors affecting the performances of rubber compounds and their processing behaviour. Mineral fillers possess considerable differences in particle geometry, depending on the crystal form of the mineral. The minimum anisometry is found with materials that form crystals with approximately equal dimensions in the three directions, which are isometric particles (i.e., calcium carbonate). The most anisometric are particles that have two dimensions much smaller than the third one (rod-shaped), or one dimension much smaller than the two others, as in the case of platelets (i.e., kaolin, talc and mica). In compounds containing fillers having identical surface area and chemical nature, but different shape, the modulus increases with increasing anisometry [9]. Particles with a high aspect ratio, such as platelets or fibrous particles (i.e., glass fibres), have a higher surface-to-volume ratio, which results in higher reinforcement of the rubber compound. The highest hardness is provided by rod-shaped or plate-like particles, which can line up parallel to one another during processing, compared to spherical particles of similar diameter.

The aspect ratio is not applied to CB and precipitated silica. The primary particles of these fillers are essentially spherical, but these spheres aggregate in such a way that the functional CB and precipitated silica fillers form clusters. The anisometry of these fillers is described in terms of *structure*, which incorporates aggregate shape, density and size. The higher the structure, the greater is the potential reinforcement.

Surface activity

Surface activity relates to the compatibility of fillers with a specific elastomer and the ability of the elastomer to adhere to the fillers themselves. If the filler particle size may be considered as a physical contribution to the reinforcement, surface activity provides the chemical contribution. The ability of fillers to react with the polymer results in chemical bonding, which significantly increases the strength.

A filler can offer high surface area, high aspect ratio and small particle size, but can still provide relatively poor reinforcement since it has low specific surface activity. CB particles, for example, have carboxyl, lactone, quinone and other organic functional groups, which promote a high affinity between rubber and fillers. Thus, the polymer–filler bonding, particularly in the case of CB, develops through active sites on the filler surface resulting in "bound rubber" attached to the filler surface. Therefore, being the bound rubber as the result of rubber–filler interactions, it can be considered as a measure of the surface activity of black fillers.

Surface activity reflects on the mechanical properties of rubber such as tensile strength, abrasion and tear resistance [10]. Both the interfacial adhesion and the formation of networks between fillers and polymer lead to the formation of high modulus compounds, which is a clear indicator that polymer–filler bonding has taken place, having the effect of reducing the mobility of polymer chains.

Surface area

The most important single factor that determines the degree of reinforcement is the development of a large polymer–filler interface that is strictly linked to the surface area of fillers. The surface area of any particle is related to its particle size as well. The smaller the particle size, the bigger is the surface area. On average, spherical particles of 1 µm in diameter have a specific surface area of 6 m^2/cm^3. This constitutes approximately the lower limit for having a significant reinforcement. The upper limit of useful surface area is of the order of 300–400 m^2/cm^3. In reality, particles have a distribution of size and are usually far from being spherical. Indeed, they may be spheroidal, cubic/prismatic, tubular, flaky or elongated. Generally, non-spherical particles can impart better reinforcement.

Particles with a planar shape have more surface area available for the contact with the rubber matrix than the isotropic particles with an equivalent particle diameter. Therefore, they show a bigger potential to reinforce the rubber compound since they results in stronger interactions between filler and rubber.

Porosity

Filler porosity can affect the properties of vulcanisates, even when its effect on reinforcement is secondary. In most cases, the pores are too small for the polymers to get inside, although some smaller molecules in the compound may do.

Surprisingly, the main characteristics of fillers (i.e., particle size, shape, surface area and surface activity) are interdependent in improving the end properties of rubber. Increasing the surface area (thus decreasing the particle size) gives higher Mooney viscosity, tensile strength, abrasion resistance, tear resistance and hysteresis, but it lowers resilience. In addition to providing the higher tensile strength required to resist to failure, small particle size, high surface area, high surface activity and high aspect ratio allow the filler particles to act as barriers to the propagation of micro-cracks. Again, an increase in the surface activity results in high modulus at the higher strain (≥300%), higher abrasion resistance, higher adsorption properties and lower hysteresis.

6.2.2 Particle-elastomeric matrix compatibility

As discussed, polymer–filler interactions are developed during the mixing process, thus influencing some characteristics of compounds, such as reinforcement, modulus, tear strength, elongation, hardness, resilience, abrasion, viscosity, heat built-up, processability and so on. These interactions depend on the following:
- external factor (i.e., the total surface area of the filler in contact with the rubber);
- internal factor (i.e., the surface activity and the chemical properties of fillers);
- geometrical factor (i.e., the structure and porosity of filler, which are minor factors) and
- strength of interactions (i.e., Van der Waals forces, polar forces, ionic forces, covalent bonding forces and steric hindrance).

Regardless of filler size and shape, the intimate contact between the elastomeric matrix and the mineral particles is essential, since air gaps represent points of permeability and zero strength (Fig. 6.2). The surface chemistry of fillers determines the affinity for the matrix, or the ability of the rubber matrix to wet the filler surface.

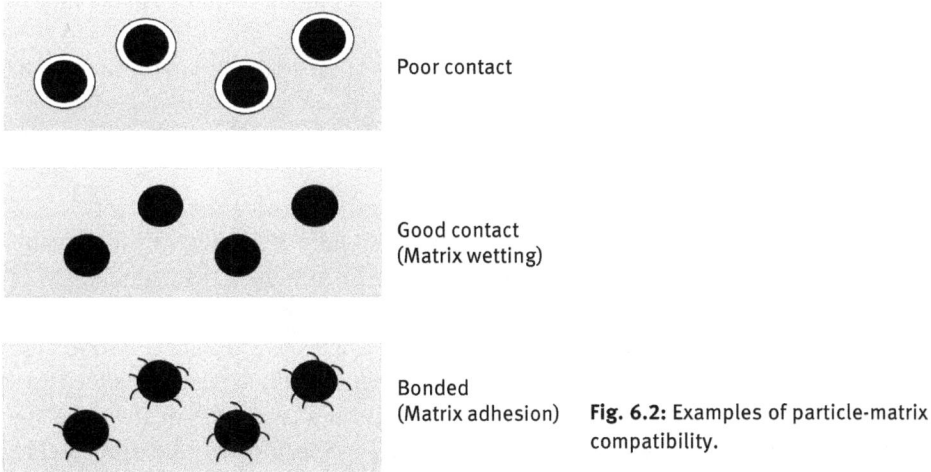

Poor contact

Good contact
(Matrix wetting)

Bonded
(Matrix adhesion)

Fig. 6.2: Examples of particle-matrix compatibility.

It is easier for most elastomers to wet the naturally hydrophobic CB surface, as compared to the naturally hydrophilic surfaces of most non-black fillers. This advantage of CB complements its reactivity.

The hydrophobicity and the reactivity of most non-black fillers can be improved with suitable surface coatings (coupling treatment). The conventional surface treatment for calcium carbonate comprises stearic acid, which improves the hydrophobicity and wettability of the filler, but does not work for the filler–matrix adhesion. Maleated polybutadiene (polybutadiene with grafted maleic anhydride functional groups) has been used as an in situ coupling agent to improve the matrix adhesion to calcium carbonate fillers. Precipitated calcium carbonate (PCC) pre-treated with maleated polybutadiene is also available.

Organosilanes are successfully employed to increase the physical properties of non-black fillers including calcium silicate, clay, mica, silica and talc [11]. The general chemical structure of organosilanes is RSiX3, where X is a hydrolysable group, such as methoxy or ethoxy, and R is a non-hydrolysable organofunctional group. The organo group may be reactive toward the rubber matrix, or it may be unreactive and serve as a hydrophobe or wetting agent.

Modification with organosilane depends on the ability to form a bond with silanol groups (–Si–OH) and/or aluminol groups (–Al–OH) on the filler surface. The hydrolysis of an alkoxysilane forms silanetriol and alcohol. Silanetriol slowly condenses to form oligomers and siloxane polymers. The –Si–OH groups of hydrolysed silane initially hydrogen bond with the OH groups on the filler surface. As the reaction proceeds, water is lost and a covalent bond is formed. The reaction of hydrolysed silane with the OH groups on the filler surface can ultimately result in the condensation of siloxane polymer, encapsulating the filler particle if a sufficient amount of silane is used. Once the filler is reacted with silane, it exposes an organophilic or organofunctional surface for the interaction with the rubber matrix, thus promoting compatibility.

6.3 Carbon black

CB is the largest volume ingredient in any black compound after the polymer itself [6, 12, 13]. It affects many characteristics of rubber as well as of the mixing cycle, thanks to the reactive organic groups on its surface that cause affinity to the material.

CBs are prepared by incomplete combustion or by thermal cracking of hydrocarbons. Although not all hydrocarbons lend themselves to economical production, essentially any combustible hydrocarbon could theoretically be used as a feedstock to produce CB. Several parameters are controlled in the process to achieve the specific characteristics of the finished CB products. Specifically, the feedstock is injected at high speed into a reactor where it is pyrolysed at high temperature (approximately 1,200–1,900 °C). The combustion reaction is controlled by quenching with water such that the oxidation remains incomplete and CB is formed.

During the combustion reaction, carbon nodules with dimensions ranging from 5 to 100 nm, depending on the grade of CB to be produced, are formed. On the basis of the proposed definitions for nanomaterials, these nodules may be considered as nano-objects. Their lifespan is very short as they immediately cluster together to form aggregates of sizes between 70 and 500 nm. They are complex constructions of randomly sized particles present in any CB in a large number; thus, the mean effect of all these individual entities must be controlled. They can have a very dense, solid structure or an open lattice-like configuration, thus determining opposite aggregate density.

After few seconds, aggregates coalesce to form agglomerates, where the bonds are electrostatic and therefore cannot be broken under normal handling conditions. The resulting agglomerates typically measure between 10 and 100 μm. The degree with which agglomeration takes place directly impacts the ability of the end user to disperse a CB in the formulation. With the single exception of conductive compounds, ultimate physical properties are achieved only with the complete dispersion of agglomerates into their constituent aggregates. In the real world, however, this desirable condition is rarely achieved. Because aggregation and agglomeration results in particle sizes well above the nanoscale, CB used as a raw material meets the definition of a nanostructured material, but does not meet the definition of nano-object.

In the CB industry, the convention refers to the smallest individually distinct unit of carbon as an aggregate and the aggregates as particles. This convention contrasts with that of other industries in which the smallest distinct unit of material is the particle. The average particle size of commercially available CB grades for rubber ranges from about 5 to 100 nm. Within this wide range, a very large number of grades exist, each one providing a unique set of properties to any compound. Distributions of particle size can be narrow or broad or even bimodal, each type influencing the characteristics of rubber. The distinction between particles and aggregates is readily detectable with the aid of electron microscopy, as shown in Fig. 6.3.

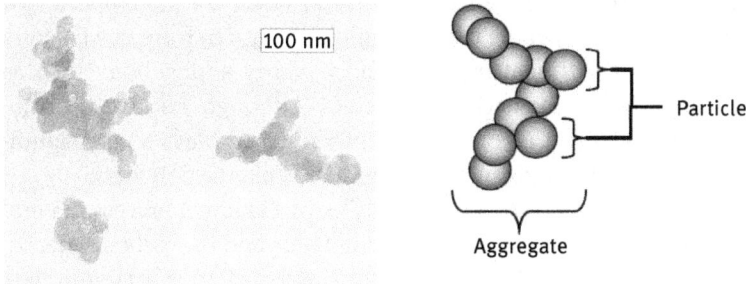

Fig. 6.3: Schematic representation of carbon black aggregates and the corresponding TEM micrograph.

Depending on the method of manufacturing, CBs are classified into four types:

- *Furnace blacks*: The largest amount of CB used today is of the furnace type. The yield process varies from 25% to 75% of the available carbon, depending upon the particle size. Details of a range of furnace blacks generally used for rubber reinforcement are listed in Tab. 6.1. They are used in every type of black-filled rubber article, thanks to their high structure and high reinforcing action. The low reinforcing furnace grades are used in carcasses of tires, and the high reinforcing ones in treads. The fine particle blacks are used where high strength and resistance to abrasion are required, that is, in conveyor belt-covers and certain types of footwear. The coarse particle blacks are used in articles such as hose, cables, footwear uppers, mechanical goods and extruded.

Tab. 6.1: Properties and characteristics of furnace blacks.

Black	Name	Surface area (m^2/g)	Average particle size (mm)
N110 (SAF)	Super abrasion furnace	140	20–25
N220 (ISAF)	Intermediate super abrasion furnace	120	24–33
N330 (HAF)	High abrasion furnace	80	28–36
N550 (FEF)	Fast extrusion furnace	45	39–55
N660 (GPF)	General-purpose furnace	37	50–60
N774 (SRF)	Semi-reinforcing furnace	28	70–96

- *Thermal blacks*: In the thermal process, oil, or more frequently, natural gas is cracked at 1,300 °C in the absence of oxygen in a hot refractory surface. The recovery is about 40–50% of the available carbon. The blacks thus obtained range in particle diameter from 120 to 500 nm. The main types of these blacks are FT (fine thermal, FT-N880) and MT (medium thermal, MT-N990). Thermal blacks are used in inner linings and inner tubes, since they provide a low degree of reinforcement and can be used at high loading. They are used also in V-belts because of their low heat build-up and in other applications such as mats, sealing compounds and mechanical goods. Although furnace CB comprises most of the worldwide CB consumption, thermal CB plays a very significant role especially in compounds utilising high-performance polymers.

- *Channel blacks*: In the now obsolete and virtually extinct channel process, natural gas is burnt in small burners with a sooty diffusion flame and the carbon is deposited by the impingement of the flame on a cool surface such as a large rolling drum or on to slowly reciprocating channel irons. The deposited black is scraped and collected. The yield from this very inefficient process is 5% or less, but very fine blacks can be made, with particle diameters ranging from 9 to 30 nm.
- *Acetylene blacks*: This is a high purity form of CB made from the thermal decomposition of acetylene gas. It is available in powder and granular forms as well as standard and high conductive grades. It provides the highest degree of structural aggregation, increased thermal and conductive properties. The low metal content meets the high standards for aerospace, automotive and battery industries. Increased absorption levels and nanoparticles size provide exceptional performance in compact applications.

Around 95% of the total CB produced is used in rubber industry, and approximately 80% of this is used in the manufacture of tires and related products. To produce the material used by the tyre industry, the CB passes through a pelletiser to compact the CB into a pellet form of millimetre size to facilitate both shipping and handling. The formation of pellets decreases the dustiness of CB, and thus reducing the exposure for workers in the rubber industry.

Similar to what occurs with precipitated amorphous silica when it is introduced into the mixer, pellets of CB are broken down during mixing, transforming back the CB into the agglomerated state. The shear forces applied during mixing break apart the agglomerates resulting in reversion to the aggregated state. As with precipitated amorphous silica, these aggregates become incorporated and chemically and/or physically bound to the rubber matrix, making them unavailable for release.

6.3.1 Nomenclature

The majority of thermal and furnace CBs are classified using a four-character naming convention as described by ASTM standard D1765 [14]. It consists of a prefix letter followed by a three-digit number. The first character is a letter that indicates the effect of the CB on the compound cure rate: N or S indicates whether the grade is a (N)ormal or a (S)low curing material. When the nomenclature system was developed, channel blacks that are slow curing were still commonly used. Afterward, a 50-fold increase in natural gas price ended the production of rubber-grade channel black. For a short period, furnace grades tried to emulate the cure characteristics of channel grades if possible. In any case, channel blacks disappeared, leaving a little need for the use of the S prefix letter. Consequently, most (if not all) currently active rubber grades carry the prefix N, as in the case of N990 where the (N) indicates a normal cure rate.

The first of the three digits is used to provide a coarse measure of the average particle size of the grade, as determined by electron microscopy and expressed in nanometres. The typical particle diameters are arbitrarily grouped, as shown in Tab. 6.1 for furnace blacks. In a very general way, both manufacturing costs and rubber reinforcement potential (i.e., abrasion resistance) increase when smaller numbers are in the first digit. The last two digits are assigned arbitrarily by ASTM D24, indicating differences within a group (structure level, modulus or any physical–chemical property) [15].

When a competitor matches a new product, it normally carries the same N number as that provided by the originator. In general, lower structure blacks have lower numbers and higher structure blacks have higher numbers. However, there are exceptions and structure values are not always proportional to the assigned second and third digits.

6.3.2 Basic chemistry and main characteristics of carbon black

CB is essentially elemental carbon in the form of extremely fine particles having an amorphous molecular structure made by micro-crystalline arrays of condensed rings [12, 16]. These arrays appear to be similar to the layered condensed rings exhibited by graphite, which is another form of carbon. Consequently, a large percentage of arrays show open edges of their layer planes at the surface level of any particle. Associated with these open edges there are many unsatisfied carbon bonds that provide the sites for the typical CB chemical activity. CB owes its reinforcing character to the colloidal morphology, size and shape of the ultimate units and the surface properties.

Although similar in micro-structure to graphite, the carbon layers in CB are less ordered. More precisely, the orientation of the micro-crystalline arrays within the amorphous mass appears to be random. This results in CB being an intrinsic semi-conductive material even if the amount of conductivity imparted to a rubber compound depends also on other factors, such as the primary particle size, structure, porosity and surface groups. Particularly, the primary particle size is the major characteristic influencing conductivity. The average width of the gaps between the particles within the aggregate is considered to be a key factor as well.

Extraneous materials, such as moisture, refractory dust, metallic oxides, coke and water-soluble salts, originating from the manufacturing process and the equipment are typically found in CB. The levels of these extraneous substances are controlled within specified limits to avoid any influence on the final rubber behaviour.

The pH of CB is mostly dependent on the condition of the water used during its production and it varies by both grade and supplier. This parameter must be considered for every rubber compound since it can affect the cure system, thus resulting in processing variations.

The five most important characteristics of CB are described a follows:

1. *Particle size*: As mentioned, particles of CB are not discrete but are fused clusters of individual particles. The fusion is especially pronounced with very fine CBs. However, the reinforcement imparted by the black is not influenced by the size of the clusters but greatly by the size of the particles within it. The average particle size of various grades of furnace CBs increases from 20 nm for N110, to 80 nm for N762 and 280 nm for N990 (Tab. 6.1). In general terms, the smaller the particle size, the poorer is the processability, the higher is the cost and the higher is the reinforcement. MT has the largest particle size of all the CBs and therefore the lowest surface area (9 m^2/g on average).

2. *Structure*: It refers to the degree of particle aggregation. A low-structure CB shows 30 particles/aggregate, while a high-structure CB has 200 particle/aggregate. The high structure comprises a high degree of bulkiness, manifested in low bulk density and high capacity to absorb oil. In rubber technology it is customary to associate high structure of fillers with high modulus of vulcanisates since the structure represents the degree to which a CB provides reinforcement to an elastomeric compound. The effect of structure is more noticeable on processing properties than on the vulcanisates properties. A black with high structure gives a high modulus rubber, not only because the CB agglomerate restricts the cross-linked network but also because the high shear forces during mixing break down these agglomerates to give active free radicals able of reacting with rubber. High structure, however, does not increase either tensile strength or tear resistance; these two properties are usually associated with reinforcement. In general, the higher the structure, the stiffer and less nervy is the unvulcanised compound and the harder is the vulcanised material. The term *nerve* relates to the elastic recovery from deformation of a raw rubber when a stress is removed from it.

   The thermal process produces blacks with little or no structure. On the other hand, the oil furnace process, using highly aromatic raw materials, gives blacks of high structure. For example, N990 MT CB is characterised by large spherical particles with very minimal aggregation. Thus, it is less reinforcing

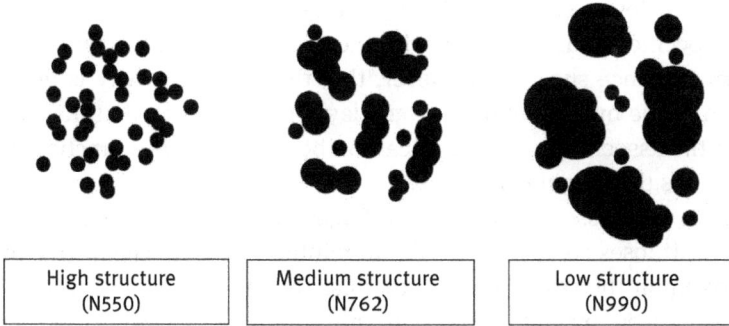

| High structure (N550) | Medium structure (N762) | Low structure (N990) |

Fig. 6.4: Examples of carbon black structures.

than even the most coarse furnace CBs, which exhibit bulky grapelike aggregates characterised by an irregular morphology (Fig. 6.4).

3. *Physical nature of particle surface*: The nature of carbon atoms on the surface of CB particles may affect the rubber reinforcement. Carbon atoms in a CB particle are arranged in layer planes. Diffracted-beam electron micrographs show that in the blacks with low reinforcing potential (thermal blacks) these layers are highly oriented. They are mainly parallel to the surface, have regular spacing and are quite large with very few defects in their network structure. On the contrary, blacks with high reinforcing potential show less orientation. Here particles are more irregular in shape, and the layers are much smaller, less frequently parallel and with more defects. This may indicate the presence of significant amounts of 'non-graphitic' carbon. Carbon atoms are relatively unreactive if they are an integral part of the layer plane. They become more reactive when they are attached to a hydrogen atom, and very reactive if present as a resonance-stabilised free radical.

4. *Chemical nature of particle surface*: CBs consist essentially of more than 90% of elemental carbon. Very small quantities of other elements such as oxygen, hydrogen and sulphur occurring in various functional groups bonded to the carbon bulk are also present. Hydrogen is distributed throughout the CB particles, whereas oxygen is confined onto the surface. Functional groups on CB surface (i.e., phenolic, ketonic and carboxylic together with lactones that are chemically combined) can react with the rubber molecules to form grafts during processing and vulcanisation. The rubber cross-linking rate is affected by the phenolic and carboxylic groups present in CB and it is increased in the presence of sulphur.

5. *Surface activity* [16]: It refers to the chemical reactivity of the CB and to its effect on the surface interaction with the polymer. It is related to the chemistry and graphitic structure of carbon. CB produced from a high purity feedstock, such as natural gas, is characterised by low surface energy and fewer surface groups,

thereby resulting in lower surface activity. Surface energy increases with an increase in the specific surface area and with the polyaromatic character of the carbon. Modern evidence suggests, however, that the surface activity is more likely a function of the number of open-edged layer planes exposed at the surface along with the associated unsatisfied carbon bonds than it is a function of the chemical groups that might exist there.

When a given CB, possessing a high reinforcement potential, is subjected to sufficient heat, it loses some of its ability to reinforce rubber. The degree of change depends upon the severity of the thermal treatment. The loss of reinforcement is evidenced by a reduction in the cure rate, modulus, tensile strength, abrasion resistance, tread wear and other physical properties. The thermal treatment induces a change in the CB morphology, since there is a loss in the surface activity. It is generally recognised that CBs with a high amount of surface activity give high reinforcement to rubbers.

6.3.3 Compounding with CB

Compounding with CB usually comes down to two fundamental decisions related to the properties required in the final vulcanisate: the choice of the CB grade and its loading [12, 13].

In many rubber carbon black has an influence second only to the polymer on compound properties and volume cost. Therefore, choosing which CB may ensure the best balance of cost and properties in a product is the key step in the compounding process of any black rubber. In the selection of the CB grade, the fundamental characteristics of CBs (particle size and structure level) must be considered, since they differentiate most grades and also influence the compound properties more than any other properties.

The logical starting point in selecting the most suitable CB is to assess the required level of reinforcement. Compound characteristics such as tensile strength, abrasion resistance, tread wear and tear resistance must be considered, bearing in mind that if these properties are improved by choosing a higher surface area, other changes can occur. Indeed, viscosity and compound cost will increase while dispersibility, rebound and other dynamic properties will be reduced. It must be recognised that the elastomer used in the compound also influence, to a larger extent, these properties. Fortunately, if CB A imparts a higher modulus in a given compound than CB B, when the polymer would be changed, black A is still expected to give the higher modulus.

In Tab. 6.2 the main effects of CB on rubber properties are summarised. Fig. 6.5 depicts the behaviour of some rubber properties depending on the CB content in rubber.

Tab. 6.2: Effect of increasing CB on rubber properties.

Property	Surface Area	Structure	Loading
Uncured properties			
Mixing temperature	Increase	Increase	Increase
Mooney viscosity	Increase	Increase	Increase
Dispersion	Decrease	Increase	Decrease
Loading capacity	Decrease	Decrease	NA
Cured properties			
300% modulus	Not meaningful	Increase	Increase
Tensile strength	Increase	Not meaningful	Increase*
Elongation	Not meaningful	Decrease	Decrease
Hardness	Increase	Increase	Increase
Tear resistance	Increase	Decrease	Increase*
Hysteresis	Increase	Not meaningful	Increase
Abrasion resistance	Increase	Not meaningful	Increase*

*Increases to a maximum, then decreases.

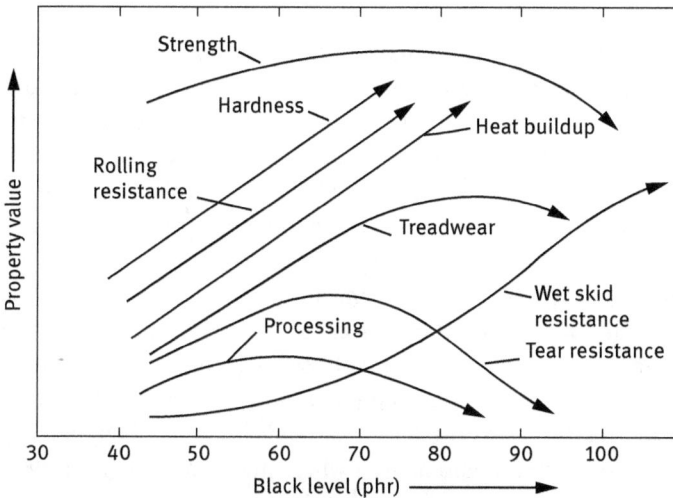

Fig. 6.5: Influence of the amount of carbon black on rubber properties.

The CB structure inhibits the elasticity of a compound in the same way as the reinforcing steel inhibits the elasticity of concrete. Aggregates with a certain shape factor (having a longer dimension) behave as very short fibres and interfere with the elastic mobility of the polymer in which they are dispersed. This stiffening effect is, therefore, more pronounced with structure than with the particle size.

Many commercial rubber compounds are developed to a given hardness specification. Since hardness varies with both the particle size and the structure, the required hardness is realised with adjustments in the CB and/or plasticiser loading. Choosing the optimum balance between the CB and the plasticiser can be a time-consuming process since nearly every compound property, as well as the cost, is affected to some extent by these two factors.

The CB grade selected for a given compound has a very large influence not only on the rubber properties but also on the methods used in compounding. It is well known that to achieve the full potential of any rubber compound, a high degree of dispersion of all ingredients in the polymer matrix must be realised. The degree of difficulty in achieving suitable dispersion worsens as both particle size and structure are decreased. For example, with thermal blacks, which have uniquely large particles, very good dispersion is readily attained in nearly any compound. In contrast, reaching the same dispersion level with N110, which contains very fine particles, can be rather difficult and may require as many as three or more passes through the mixer with adequate cooling between.

6.4 Non-black fillers

Clays represent the largest volume of non-black filler used in rubber industry because of their cost effectiveness in terms of providing beneficial reinforcing and processing properties at a modest price [3, 7]. They are second to CB in this respect.

Clays can be formulated to high loadings in most elastomers. Viscosity builds moderately with the clay loading, but formulations with 150–200 phr are reported to be feasible [17]. The main factor to be considered in adding clay to most formulations is the reduction in the cure rate. This reduction requires the addition of an activator and/or an increase in the dosage of accelerator.

Rubber filler clays are classified as *hard* or *soft* based on their particle size and the stiffening effect in rubber. *Hard* clay is very fine grained and has an average particle size from 250 to 500 nm. It will impart high modulus, high tensile strength, stiffness and good abrasion resistance to rubber compounds. *Soft* clay has an average particle size from 1,000 to 2,000 nm and is used where high loadings (for reducing cost) and faster extrusion rates are more important than strength.

More *hard* clay than *soft* is used in rubber because of its semi-reinforcing effect and its utility as a low-cost complement with respect to other fillers. In certain compounds it is used as a substitute for a portion of the more expensive CB, without sacrificing the physical properties.

The most important clays are kaolin (aluminium silicate) and its derivative and the amorphous precipitated silica (see Section 6.5).

Recently, layered silicates have attracted a great deal of interest as nanocomposite reinforcements in rubbers owing to their intrinsically anisotropic character, swelling capabilities, high aspect ratio and the plate-like morphology. Besides precipitated silica, layered silicates and kaolin, some other white fillers are worthy of mention, not because of their reinforcement qualities but because of their high consumption. They include calcium carbonate, mica (potassium aluminium silicate), talc (magnesium silicate) and titanium dioxide.

Calcium carbonates for rubber, often referred to as *whiting*, fall into two general classifications. The first is the wet- or dry-ground calcium carbonate, spanning the average particle sizes from 700 to 5,000 nm. The second is the PCC with fine and ultrafine products, extending the average particle size range down to 50 nm. The much smaller size of PCCs provides a corresponding increase in the surface area. The ultrafine PCC products (<100 nm) can provide surface areas equivalent to the hard clays. Manipulation of manufacturing conditions allows the production of PCCs of several distinct particle shapes. Precipitated calcite, with isometric prismatic particles, is the form generally used in rubber compounding. In general, dry-ground calcium carbonate is probably the least expensive compounding material available, and more can be loaded into rubber than any other filler. Water-ground calcium carbonate is slightly more expensive, but offers better uniformity and finer particle size. The ground natural products used in rubber have low aspect ratio, low surface area and low surface activity. They are widely used, nevertheless, because of their low cost and because they can be used at very high loadings with a small reduction of the compound elongation or resilience. This follows from the relatively poor polymer-filler adhesion potential.

Mica is a chemically inert phyllosilicate, resistant to thermal decomposition. Due to its platy nature, it imparts excellent flexural and bridging characteristics, which contribute to the crack resistance in some applications. The platy nature also enhances the release performance in rubber moulding and sheet production.

Talcs are used as semi-reinforcing fillers in all rubber applications. Because talc particles are organophilic, they show a marked affinity for elastomers, resulting in improved filler–elastomer cohesion. This enhances tensile strength, elongation, modulus and tear resistance of the vulcanisates.

Titanium dioxide finds extensive use in white products such as white sidewalls of tyres where the appearance is more important than the reinforcing effect.

6.5 Amorphous precipitated silica

Amorphous precipitated silica is considered semi-reinforcing white fillers and its structure is almost similar to that of CB, consisting of aggregates [3, 5, 8]. The world production of amorphous precipitated silica is 1.3 million tons, where one-third is used in tire production, especially in the tread, to reduce the fuel consumption of vehicles, thus contributing to a decrease in the emissions of greenhouse gases.

Amorphous precipitated silica is produced by a solution process by a reaction between alkali silicate solutions and either concentrated sulphuric, hydrochloric or carbonic acid under controlled conditions, followed by precipitation [11]. Reaction conditions are manipulated based on the required particle size. During precipitation, there is an instantaneous formation of primary nanoparticles size (from 2 to 40 nm) of a very short lifespan. These particles immediately cluster to form aggregates (from 100 to 500 nm in size) based on covalent bonds. Because of the nature of these bonds, these aggregates cannot disaggregate under standard conditions. Subsequently, the aggregates bond together by hydrogen bonding to form agglomerates from 1 to 40 μm. Because of the presence of both aggregates and agglomerates, precipitated amorphous silica meets the ISO definition of a nanostructured material, but it does not meet the definition of nano-object.

The reinforcing properties of precipitated silica can usually be related to the particle size: 10–40 nm particles are reinforcing, while >40 nm particles are semi-reinforcing. Because of the difficulty in measuring the size of very small particles, as with CB, the surface area is usually used for classifying the various grades of amorphous precipitated silica. For example, silica having surface area in the range of 125–250 m^2/g is generally reinforcing, while products in the range of 35–100 m^2/g are semi-reinforcing.

Other properties of silica and silicates influence their use in rubber compounds, such as the extent of hydration, pH, chemical composition and oil absorption, but they are of secondary importance.

The addition of silica to a rubber compound offers a number of advantages such as the improvement in tear strength, resistance to flex fatigue, abrasion and hot tear resistance, reduction in heat build-up and increase in the compound adhesion in multi-component products, such as tyres.

Low-moisture-precipitated silica is practically not possible to manufacture because of high cost and also the natural tendency of the silica molecules to absorb or loose moisture to maintain the equilibrium with the relative humidity of the environment. Precipitated silica is usually sold with about 6% adsorbed free water and a surface essentially saturated with silanol groups (–SiOH) readily formed by hydrolysis of the surface silica groups in the presence of moisture. Silanol groups behave as acids (–SiO–H+) and are chemically active. While this would normally be considered an advantage for a rubber filler, it can cause problems with other compounding ingredients, such as the cure system.

Silanols show similarities to carboxylic acid groups in their reactions with amines, alcohols and metal ions. Most of the accelerators used in sulphur cure systems contain an amine group. Strong adsorption or reaction with the filler particles can decrease the amount of accelerator available for the vulcanisation reactions. This can give slower cure rates and a reduced state of cure. Similar effects can result from the reaction of soluble zinc ions with silica particles. The adsorption or reaction of accelerators by hard clay usually requires about a 15–25% increase in acceleration.

Free water also inhibits the bonding between the rubber matrix and the silica particles. To overcome this drawback, the silica surface is modified by coupling agents, like high-molecular weight polyethylene glycol, glycerine, siloxanes, triethanol amine and so on, that insulate the surface of the silica particles from the reaction with the accelerators and the soluble zinc. Various silanes are also used as coupling agents that react with the silanolic OH groups forming a graft with the polymer (see Section 6.2.2). Since the silane-coupling agent provides a means to bond the clay particles to the rubber, increased modulus, tear strength, tensile strength, adhesion in multicomponent products and improved ageing properties are obtained. Mercaptosilane is usually the most cost-effective choice and the treatment is typically made in situ, with silane added to the mixer after silica and before other additives that can interfere with the silica–silane reaction.

Generally, the amorphous precipitated silica, when compared to CBs of the same particle size, does not provide the same level of reinforcement. However, the deficiency of silica largely disappears when coupling agents are used with it.

6.6 Anti-degradants

The presence of unsaturation (C=C) in the backbone structure makes the elastomeric chains susceptible and vulnerable to be attacked by light, oxygen, ozone and also by thermal degradation and more generally by natural weathering [1, 7].

At macro-molecular level, the loss in physical properties associated with both natural and accelerated ageing processes is normally caused by either chain scission, resulting in the reduction of chain length and the average molecular weight, cross-linking resulting in a three-dimensional structure and higher molecular weight and chemical alteration of the molecule by the introduction of new chemical groups. Nonetheless, the altered network now contains increased chain-end defects; thus, strength and elongation are reduced.

Natural rubber, polyisoprene and butyl rubbers degrade predominantly by chain scission, resulting in a weak, sticky, softened material often showing surface tackiness. Chemical analysis shows the presence of aldehyde, ketone, alcohol and ether groups resulting from the oxidative attack and mostly at α-hydrogens and at double bonds. SBR, neoprene, EPDM, polybutadiene and acrylonitrile degrade by cross-linking, giving brittle compounds with poor flexibility and elongation. For some compounds in the early stages of oxidation, there is an equality in the extent of chain scission and cross-linking such that modulus does not change.

All rubbery materials, whether natural or synthetic, cured or uncured, contain a certain amount of chemical unsaturation, subjected to degradation. Consequently, anti-degradants are added to retard or prevent the polymer breakdown, to improve the ageing resistance and to extend the service life of vulcanisates. Anti-degradants can react with the degradation agents or can interfere with the reaction mechanisms that otherwise would culminate in the rubber degradation.

The selection criteria governing the choice and the use of anti-degradants are as follows:

- Discoloration and staining: For elastomers containing CB more active amine anti-oxidants are preferred.
- Volatility: As a rule, the higher the molecular weight of the anti-oxidant, the less volatile it will be. Hindered phenols tend to be highly volatile compared with amines of equivalent molecular weight. Thus, the correct addition of anti-oxidants in the compound is critical for avoiding any loss of material.
- Solubility: The low solubility of an anti-oxidant causes blooming onto the surface with consequent loss of protection in the vulcanisates. Therefore, the solubility of anti-degradants, particularly anti-ozonants, controls their effectiveness. They should not be extracted by water or other solvents, such as hydraulic fluid, during their service life.
- Chemical stability: Stability of anti-degradants against heat, light, oxygen and solvents is required for durability.
- Concentration: Most anti-degradants have an optimum concentration for maximum effectiveness after which the material solubility becomes a limiting factor. Para-phenylenediamine (PPD) offers good oxidation resistance at 0.5–1 phr and anti-ozonant protection in the range 2–5 phr. Above 5 phr PPD tends to bloom.
- End use of the rubber (static or dynamic applications).

6.6.1 Anti-oxidants

A concentration of only 1–2% of oxygen is normally sufficient to cause severe deterioration in any elastomer [6–8]. The principal mechanism of oxygen attack involves an autocatalytic free radical reaction. The first step is the creation of macro-radicals as a result of hydrogen abstraction from the rubber chains by a proton acceptor. Oxidation continues by reaction of the macro-radicals with oxygen and the subsequent formation of peroxy and hydroperoxides radicals. Oxidation is accelerated by heat and UV light and by the presence of some metals, such as copper, cobalt and manganese. For sulphur-cured vulcanisates the oxidation rate increases with the content of sulphur.

A proper anti-oxidant slows down the oxidation process by interrupting the degradation reactions either by capturing the free radicals formed and/or by ensuring that the formed peroxides and hydroperoxides decompose into harmless fragments, without degrading the polymer and without initiating new free radicals able of propagating the chain reactions.

Anti-oxidants are categorised into two classes. The first type, called preventive anti-oxidants, reacts with hydroperoxides to form harmless, non-radical products.

The second type, named chain-breaking anti-oxidants, destroys the peroxy radicals that would otherwise propagate.

The majority of commercially available anti-oxidants belong to two main chemical classes: amines and phenolics, which represent staining and non-staining types, respectively. Anti-oxidants derived from p-phenylenediamine and diphenylamine are highly effective scavenger of peroxy radicals. They are more effective than the phenolic anti-oxidants for the stabilisation of easily oxidisable unsaturated elastomers. Phenylene diamine derivatives are used primarily for elastomers containing CB because of their intense staining effects. Substituted diphenylamines are especially useful in neoprene type. They tend to show a directional improvement in the compound fatigue resistance. The non-staining anti-oxidants include hindered phenols, hindered bisphenols and hydroquinones. Hindering the phenolic hydroxyl group with at least one bulky alkyl group in the ortho position appears necessary for having high anti-oxidant activity. Steric hindrance decreases the ability of a phenoxy radical to abstract a hydrogen atom from the substrate and to produce an alkyl radical capable of initiating oxidation. However, because of their low molecular weight hindered phenols tend to be volatile.

6.6.2 Anti-ozonants

Ozone, even when present in the atmosphere at only a few parts per hundred million, readily cleaves the C–C bonds in elastomers [6–8]. As a result, an unsaturated rubber, exposed to ozone under stressed conditions, quickly develops surface cracks perpendicular to the stress direction. The severity of cracking increases rapidly if a strain of the order of 10% is applied.

PPDs are the only class of anti-ozonants used in significant quantities for stopping the ozone cracking in diene rubbers, thus improving their resistance to fatigue, oxygen, heat and metal ions. There are three general categories of PPDs used as anti-ozonants:

1. Dialkyl PPDs: They induce higher levels of scorch than other classes of PPDs. They migrate faster than other classes due to their low molecular weight.
2. Alkyl–aryl PPDs: They show good dynamic protection, good static protection when combined with wax, better processing and scorch safety and slower migratory properties.
3. Diaryl PPDs: These are less active than alkyl–aryl PPDs and have a tendency to bloom.

Waxes are an additional class of materials used to improve the ozone protection of rubber primarily under static conditions (i.e., storage of tyres in a warehouse). Wax protects rubber against the static ozonolysis by forming a barrier onto the surface, since it continuously migrates from the bulk of the rubber, maintaining an equilibrium concentration at the surface. Micro-crystalline wax migrate to the rubber surface at a slower rate than paraffin wax and performs better at high service temperature, whereas paraffin waxes protect best at low temperatures.

6.7 Softeners and lubricants

Softeners include a wide variety of oils and synthetic organic materials, which do not react chemically with rubbers but serve primarily as processing aids, thanks to their plasticising function [7]. Indeed, they are called also plasticisers. They are used for a number of reasons:
- to decrease the viscosity and thereby to improve the workability of the compound;
- to reduce both mixing temperature and power consumption;
- to reduce hardness;
- to reduce the low-temperature brittle point;
- to aid in the fillers dispersion;
- to reduce both mill and calender shrinkage and
- to provide lubrication in extrusion and moulding.

The proper lubricant selection is important. The oil must be non-toxic and compatible with the rubber and the other compounding ingredients used in the formulation, since incompatibility will result in poor processing characteristics and/or bleeding in the final products. Again, they have to act at low-dosage level.

The most important class of softeners comprises the hydrocarbon oils distinguished in three primary categories: paraffinic, naphthenic and aromatic. All the three classes of oils are used at 2–10 phr and contain high levels of cyclic carbon structures. Differences are in the number of saturated and unsaturated rings. For general all-round properties, naphthenics are preferred. Certain esters of organic acids or phosphoric acid are used as plasticisers when petroleum oils may be unsuitable (i.e., because of the incompatibility with the polymer), as in the case of NBR and CR. Examples are dibutyl phthalate, dioctyl phthalate, polypropylene adipate, trixylyl phosphate (TXP) and so on.

Process oils used in the rubber industry to improve the processability are called processing aids. They are important because of they increase both the efficiency and productivity, but reduce the energy consumption. Thus, this decreases the production costs, but maintaining high quality of the product.

Processing aids reduce both viscosity and elasticity of rubber using two different mechanisms:
1. Lowering of the polymer molecular weight, thanks to the molecular entanglement reduction and the subsequent easier flow of polymeric chains.

2. Lowering of the intermolecular interactions, helping the macro-molecular chains to flow more easily.

In both cases, there is no reduction in molecular weight; hence, the final properties of vulcanised rubbers are unaffected.

The efficiency of processing aids depends on their degree of miscibility and/or solubility with the polymer. When the oil is completely miscible (or soluble) with the polymer, it is added in small amounts, leading to the best efficiency during processing, since it is most effective in reducing the system's viscosity. If the miscibility is partial, higher amount of oil is required, but in any case, the efficiency is reduced, since the oil is less effective in diminishing the viscosity. This behaviour worsens when oil and polymer are immiscible (insoluble): by adding larger amount of oils, the efficiency decreases.

References

[1] Dick, JS. Rubber technology – Compounding and testing for performance. Hanser Publishers, Munich, Germany, 2001.
[2] Kraus, G. Reinforcement of elastomer. Wiley-Inter science, New York, USA, 1965.
[3] Katz, HS., Milewski JV. Handbook of fillers and plastics. Van Nostrand Reinhold Co., New York, USA, 1987.
[4] Donnet, JB. Nano and microcomposites of polymers elastomers and their reinforcement. Compos. Sci. Technol. 2003, 63, 1085.
[5] Rothon, R. Particulate-filled polymer composites. Longman Scientific&Technical, New York, USA, 1995.
[6] Mark, JE., Erman, B., Eirich, RF. Science and technology of rubber. Eds., Academic Press, New York, USA, 1994.
[7] Blow, CM. Rubber technology and manufacture. The Chemical Rubber Co., Ohio, USA, 1971.
[8] Franta, I. Elastomers and rubber compounding materials. Elsevier, New York, USA, 1989.
[9] Wang, MJ. Effect of polymer-filler and filler-filler interactions on dynamic properties of filled vulcanizates. Rubber Chem. Technol. 1998, 71, 520.
[10] Wolff, S. Chemical aspects of rubber reinforcement by fillers. Rubber Chem. Technol. 1996, 69, 325.
[11] Morton, M. Rubber technology. Van Nostrand Reinhold Co., New York, USA, 1987.
[12] Donnet, JB, Bansal, RC, Wang, MJ. Carbon black: science and technology. CRC Press, Boca Raton, USA, 1993.
[13] Donnet, J, Bansal, R, Wang, MJ. Carbon Black. Marcel Dekker Inc., New York, USA, 1993.
[14] ASTM D1765-10, Standard Classification System for Carbon Blacks Used in Rubber Products.
[15] ASTM D24, Carbon black.
[16] Darmstadt, H, Cao, NZ, Pantea, DM, Roy, C, Sümmchen, L, Roland, U, Donnet, JB, Wang, TK, Peng CH, Donnelly, PJ. Surface Activity and Chemistry of Thermal Carbon Blacks. Rubber Chem. Technol. 2001, 73, 293.
[17] Bateman, L. Chemistry and physics of rubber-like substances. John Wiley & Sons, London, UK, 1963.

7 Rubber bonding

7.1 Introduction

Bonding rubber to a given substrate is a generic phrase covering a number of inter-dependent processes. Currently, specific rubber bonding agents are used to connect elastomers to metal (i.e., iron, steel, aluminium and brass) or to a wide range of thermoplastics (i.e., polyamides, polycarbonate and polyoxymethylene) as well as to glass and textiles. Twenty years ago, the subject was regarded as a *black art*. Now the technology allows the production of uniform high-quality products free from failure. Much of the science behind the rubber-bonding technology remains in the uncertainty since it is based on experience, but it will become an important part of future developments.

Rubber-bonded items are used in different fields, as that of isolation of noise and vibration in automotive, in many engineering applications, such as to decouple the translational movement for bridges and buildings and so on. The strength of the provided bond is higher than the cohesive energy of the rubber alone. Thus, it is quite usual that when the bonded part is exposed to stress tests, the elastomer breaks, but not the bond.

As well known, through compounding, the elastomer's performances can be enhanced, but no single elastomer can be compounded to meet all applications. In the same manner, no single adhesive and/or bonding agent can provide the needed levels of adhesion and environmental resistance to all rubbers. Therefore, when bonding a particular compound to a given substrate, the adhesive choice can vary depending upon the cure system, the environmental application of the bonded assembly, the substrate to which the rubber is going to be bonded, the moulding method and the geometry of the part. Other factors affecting the best adhesive selection might include colour, conductivity and means of application.

7.2 Basics of rubber bonding

Three essential elements form the core of the bonding process: the rubber, the sub-strate and the bonding agent.

The selection of the compound mainly depends on the final vulcanisate specifi-cations. For example, the rubber for a highly engineered automobile engine mount will be selected for its dynamic performance in controlling vibrations and for its ability to endure under-the-hood operating conditions. Conversely, the elastomer for an engine seal must provide a terrific resistance to the attack by engine fluids.

All the compounding ingredients (i.e., vulcanising system, fillers, extender oils/plasticisers, anti-degradants, etc.) affect the 'bondability' to an higher degree

https://doi.org/10.1515/9783110640328-007

in the case of the non-polar diene elastomers (EPDM, IIR and NR), and to a lesser extent, in the case of more polar types, such as CR and NBR, that are easier to bond. The amount of sulphur in the compound has a significant role on bonding efficiency: sulphur levels of 1 phr or higher have a favourable effect on bondability, whereas little or no sulphur results in a compound that is more difficult to bond. Of the more commonly used accelerators, MBT generally allows good bondability. ZDMC and the ultra-accelerators such as TMTD detract from bondability, particularly in EV or semi-EV cure systems. Elastomers not cured with sulphur systems are easier to bond through the inclusion of high surface area fillers. They become more difficult to bond when compounded with certain oils, plasticisers and waxes.

The type and amount of filler is critical as well. Compounds containing from 40 to 80 phr of carbon black are easier to bond than those with lower black levels. Non-black fillers, such as clays and silica, also facilitate bonding.

Waxy or oily compounding ingredients that migrate to the vulcanising elastomer surface cause bonding difficulties. These include low molecular weight polyolefin auxiliaries, (i.e., low-melting polyethylene and polypropylene in processing aids/lubricants), aromatic oils and fatty acid esters. Naphthenic or paraffinic oils are less problematic. High levels of anti-ozonants and certain anti-oxidants, particularly the p-phenylene diamine type, may detract from bondability.

The overall effects of elastomer blending can impact on bondability as well, thus limiting the adhesive selection. For example, blends of NBR and NR will be more difficult to bond than compounds comprised entirely of nitrile elastomers.

The substrate is chosen by considering the final use of the bonded system and to the life cycle of the rubber item. Almost any material can be bonded to rubber, if it can withstand the heat and pressures of the rubber-moulding process. The most traditional substrate is steel in all its forms and grades. There is an increasing use of aluminium alloys and polyamides to save weight. For practical purposes, polyolefin plastics show a limited use. If required, some area of the substrate can be masked just to avoid there the application of the bonding system.

The right adhesive system for a given application must provide an excellent bond with rubber under the specified vulcanisation conditions, as well as it must maintain the bond under the service conditions and during the vulcanisate life. The adhesive systems used for bonding rubber during the vulcanisation process are polymer solutions having a paint-like consistency. They consist of a one- (self-priming) or two-coat system, which works in the same way. The majority of adhesives used today employ a primer on the substrate to be bonded, followed by a topcoat adhesive (two-coat systems). Primers consist of polymers and phenolic resins that promote the adhesion not only to the substrate but also to the topcoat adhesive. They may contain metal oxides, carbon black and other fillers that may help in bonding. Topcoat adhesives are generally designed to bond a certain type of elastomer or a range of elastomers. They consist of mostly halogenated polymers with specific additives or organosilanes that act as a carrier and film former for the specific bonding agent included inside it. The primer is generally of a different colour than the topcoat to distinguish between them. Traditionally the primer is grey and the topcoat black, but other colour combinations are also available.

The bonding system is generally applied by brushing, spraying or dipping processes to reach the proper dry film thickness and to ensure the adequate adhesion between the rubber and the substrate. The layer thickness varies based on the nature of the rubber formulation and the coating process itself.

After the application of the adhesive system, the coated substrate is placed in a mould ready for the moulding cycle, where the raw rubber is vulcanised in the proper range of temperature. The time is dependent upon the mass of the item to be moulded and the cure rate of the specific elastomer. The bonding between the rubber and the adhesive-coated substrate takes place during the vulcanisation process when the primer cures as well. The topcoat adhesive, that is, the bonding agent, cross-links with both the elastomer backbone and the primer. Bond formation appears to be associated with the development of a very high modulus layer in the rubber immediately adjacent to the surface of the substrate. Primers bond to metal surfaces via chemisorption, forming covalent bonds with them. Like the primer, the topcoat must cure during the vulcanisation cycle. These types of adhesives generally require more than 130 °C as activation temperature.

7.3 Selecting the bonding agents

The adhesive selection is primarily based upon the type of rubber being bonded. However, sometimes the type of substrate is also a factor. For example, when bonding styrene–butadiene rubber (SBR) to a metal, the primer might be suitable (two-coat system), whereas if the same SBR has to be bonded to a polyester textile, a one-coat adhesive must be used. Usually adhesive suppliers have selector guides available for helping in the best choice.

Generally, any bonding agent should be easy to use and elastomer compatible, thus providing bonds characterised by high strength and durability under the expected service conditions. Depending on the selected compound and the specified vulcanisation process, one or more adhesive may offer this compatibility, especially in the case of adhesives that bond natural rubber and most of the sulphur vulcanisable synthetic diene elastomers.

Many of the high-performance elastomers (i.e., peroxide-curing silicone elastomers, ionically curing fluoroelastomers and rubber not vulcanised by sulphur) require adhesive bonding chemistries that are different from those belonging to the broad-purpose formulations. These adhesives for specialty elastomers, containing organosilane, often are applied in one coat.

The adhesive selection must consider the design of the part, the moulding method and the pre-baking tolerance. The design or geometry of the assembly influences the bonding and how well that part will withstand the service environments. For example, fluid engine mounts or bushings (i.e., those with contained fluid) may place atypical demands on the environmental resistance of the cured adhesive. If the elastomer–metal interface is exposed to a confined fluid, such as a hot glycol/water mixture, the adhesive system will need to withstand this particular service exposure.

The moulding method affects the tendency for undesirable wiping or sweeping of the adhesive. This phenomenon sometimes results when a molten elastomer compound moves across the adhesive-coated metal part surfaces prior to vulcanisation. Under these conditions, some adhesives can be swept away from the interfaces where they are needed.

As mentioned, bonding is dependent on the chemical reactions that occur at elevated temperatures between the adhesive and the vulcanising compound across the adhesive–elastomer interface. If these chemical reactions begin in the adhesive before any contact with the elastomer, a significant amount of the adhesive-bonding capability can be lost. In other words, the reaction of some key adhesive ingredients from the adhesive-coated substrate surface can cause the adhesive to lose some of its bonding activity. Therefore, it is essential to know the pre-baking tolerance of any bonding agent, that is, the adhesive's ability to withstand high-temperature exposure before coming in contact with the vulcanising rubber.

Adhesives for rubber-to-metal bonding were first developed using solvents such as alcohols, ketones, halogenated and aromatics, which lead to viscosities suitable for brush applications. However, many adhesives contain aromatic solvents, such as xylene, together with lead compounds or other heavy metals.

Currently, adhesives that do not use solvent as the carrier nor do they contain any heavy metals are available. These aqueous-based adhesives, composed of the primer and the topcoat in emulsion form, provide the same effective adhesion as the solvent-based systems. The application methods for water-based bonding agents are similar to those for the solvent-based systems, but inserts do need to be pre-heated before spraying with the primer, and reheated before applying the topcoat. Drying times are surprisingly quick, and so no barrier to high volume production exists. Unfortunately, the bonding agent manufacturers had to overcome some difficulties in the production of aqueous bonding agents, since the preparation of active polymers in stable emulsion form is not easy. For this reason, the chemical structure of the backbone polymers is not the same as for that of the solvent-based agents and also the cross-linking agents are different. In addition, it is necessary to prepare many specific grades for various chemical structures of elastomers for reaching adhesive performances as high as those of solvent-based agents. Finally, the application process of the aqueous agents onto the substrate to obtain a continuous layer with good film properties requires good experience, but the whole process cost is high.

7.4 The bonding process

The following three steps may be highlighted in every bonding process:
1. Substrate preparation
2. Adhesive application
3. Moulding, curing and finishing

7.4.1 Substrate preparation

A proper surface preparation is essential to achieve the maximum bond strength. Tab. 7.1 summarises the appropriate surface-cleaning procedures together with some guidelines and recommendations for metallic and non-metallic surfaces.

Tab. 7.1: Surface preparation methods.

Substrate	Mechanical treatment	Chemical treatment
	Metals	
Low carbon steel	Blasting with clean, sharp 40 mesh steel grit, sand or aluminium oxide grit	Phosphate
Stainless steel Aluminium Brass and copper	Blasting with clean, sharp 40 mesh sand or aluminium oxide grit	Acid etching Chromate conversion Ammonium persulphate etching
Zinc and cadmium		Phosphate or chromic acid
Titanium		Fluoridric acid pickle
	Plastics	
Polyamide Cured phenolic Epoxies Polycarbonate	Light blasting using sand or aluminium oxide grit	Cleaning with alkaline cleaner or solvent
Polytetrafluoro ethylene (Teflon; PTFE)	NA	Sodium–naphthalene etching
Polyoxymethylene	NA	Chemical etching

The foundation of a good bond is a clean and stable substrate surface. Therefore, the first key step in the preparative process for the inserts is their cleaning. Any protective oil, solid lubricant or grease layer that may be present must be removed by degreasing in perchloroethylene, trichloroethylene or 1,1,1-trichloroethane to ensure that the bonding agent achieves a good contact and an optimal wetting of the substrate surface.

Today, aqueous degreasing systems can replace the traditional solvents and are highly effective methods for surface preparation before the grit blasting. More precisely, the alkaline cleaning before and after blasting is preferred. The alkaline cleaner is a mixture of surfactant, builder and sodium or potassium hydroxide. The builders are combinations of sodium carbonate, sodium silicate, borates or sodium gluconate. The surfactant action is very important in assisting the degreasing because of the encapsulation of the oil or grease residues by the hydrophobic groups of the surfactant. The cleaner can be used by dipping in heated baths (50–100 °C), sometimes with additional ultrasonic activation, and by spraying at 60 °C. The alkaline bath must have time, temperature and concentration controls and an overflow system. A final efficient rinsing with cold and hot water is absolutely necessary in all cases to ensure the removal of any alkali and detergent traces.

In the case of metallic substrates, any scale, rust or other oxide coatings, which are insoluble materials, must be removed from the surface by mechanical or chemical treatments. Since oxidation may have a negative effect on the bond, the metal should not be stored for long after tumbling.

Mechanical treatments for the substrate preparation include blasting, abrasion, machining or grinding, where clean grit or abrasives must be used, depending on the metal. These methods remove dry soil and corrosion, increase the surface area and provide an active surface for bonding. The chemical treatment is slightly more complex and again depends upon the metal to be treated.

For plastic surface, a simple cleaning is usually sufficient. Treatment solutions as well as the rinse water and the drying air purity must be controlled. However, polyoxymethylene and PTFE must be treated chemically. A number of phosphate processes are available for the superficial etching of plastics, with some characteristics in common. They produce consistent coating in the range of 1.5–3.5 g/m^3 that lead to excellent bonds. Polyacetal inserts can be bonded but require careful etching and rubber moulding temperatures below 150 °C. PTFE provides a useful low friction material for use in anti-roll bar bushes. It can be bonded successfully to rubber by chemically etching the insert surface of the plastic prior to the application of the bonding agents.

The cleaned and prepared substrates must be stored, preventing any exposure to dust, moisture, chemical fumes, mould sprays and other contaminants. In any case, the primer should be applied as soon as possible after the surface preparation. For example, aluminium should be primed the same day.

Irrespective of the process route that is followed in the substrate preparation, the emphasis has to be on its control to ensure consistency of results. Degreaser plants need to be regularly monitored to minimise any build-up of contaminants or change in pH. Since the grade of grit used in shot-blast operations is important and affect the environmental resistance of the finished item, the profiling of particle sizes is a useful tool to monitor the machine effectiveness and to ensure that the dust levels remain low. Differences in surface profile due to changes in the size range of the grit will not cause bond failure, but the presence of dust and debris in the grit will. Dust will adhere to the newly cleaned metal surfaces and it will be very difficult to remove, even with a second degreasing operation.

7.4.2 Application of the bonding system

The method selected for the application of the bonding system usually depends upon the shape of the part and how many parts are to be coated in a given time period. For example, in the seal industry, hundreds of small metal rings are coated at one time, sometimes for three shifts a day. However, for a large earthquake-bearing pad or the inside of a solid rocket booster motor for the space shuttle program, only a few parts a week might be coated. Any conventional coating method can be used to apply the bonding agent, such as spray gun, brush, dipping or roller. The methods for applying bonding agents are subject to continual review just to improve the efficiency.

Dipping is used for solvent and aqueous adhesives. It accommodates both large and small production runs, depending on the level of automation especially for automotive applications. If the bonding agent is a dispersion, it is not ideal for dipping.

Spray gun application (automatic multi-nozzle equipment) is the most practical method, because it provides the highest level of bond performance and the most rapid evaporation rate of solvent for any application. Brushing is recommended only for solvent adhesives and it is useful for small runs or production that is not continuous, whereas roll coating provides an excellent method of coating large flat areas as well as cylindrical objects.

Many adhesives require stirring prior to the application to ensure a complete homogeneity. Slow agitation is suggested so as not to cause foaming. For small containers, hand mixing is adequate, but for larger size containers an air-powered mixer is suggested. If the bonding agent is not well mixed, it can result in bond failures since it may not be present in sufficient amount on the treated surface. Some adhesives are already in the solution (no dispersed materials), and so do not require mixing.

The adhesive for spraying or dipping has to be diluted. The technical literature on each product should be consulted to determine the proper diluents to be used and to determine what amount to add to achieve the necessary viscosity for the proper application.

The key factor in the adhesive application is the right amount of product put on the substrate surface. Insufficient film thickness causes bond failures and poor environmental resistance. Excess film thickness can cause runs/tears in the film, which in turn entraps the solvent resulting in a bond failure. Primers are generally applied in the range of 5–10 μm dry. The topcoat adhesive is applied in the range of

15–25 µm dry. One-coat adhesives are applied around 25 µm dry. When bonding specialty elastomers such as silicones and fluoroelastomers, the one-coat adhesives are applied as very thin films, less than 1 µm.

After coating, the parts may be stored for some time depending upon the adhesive type. Most general-purpose bonding agents can be stored up to 30 days, while certain specialty adhesives may be limited to less than 3 days. The adhesive-coated parts are generally protected by the use of desiccants in the container with the parts. The supplier product literature should be consulted and confirmed by practical experience.

7.4.3 Rubber moulding and finishing

Moulding is the most critical step in the bonding procedure. Any variation in the individual moulding parameters can result in bond failures or a high scrap rate. Generally, the most preferred method of moulding is injection, since it allows the greatest control over the whole process, including the rubber bonding.

When designing the mould, provisions for an easy loading of the adhesive-coated inserts as well as for the easy removal of the vulcanised part must be assessed. Again, moulds need to be designed to ensure the exact balance between cavities and the elimination of trapped gases. The presence of gases causes a high incidence of bond failures through the 'diesel effect', whereby elements of the bonding agent film can burn under the combined effects of heat and high-pressure gas. Lack of balance between cavities will result in some components that are imperfectly formed and give rise to bonds that may fail.

If a coated insert is left sitting in a hot mould cavity for too long prior to coming in contact with the elastomer to be bonded, the adhesive may pre-react, thus losing its bonding ability. Every adhesive has its own prebake resistance. In addition, the type of rubber to be bonded affects the pre-bake resistance. Most adhesives show at least 5 min of pre-bake resistance at 160 °C. Generally, this time is enough to load all the inserts and to begin moulding. Higher temperatures can reduce the pre-bake resistance.

Since some adhesives have poor hot tear resistance at the end of vulcanisation, a bond failure can occur during demoulding. For overcoming this drawback, sometime mould releases are used on moulds to allow the vulcanisate to be released at the end of the cycle, preventing the rubber from getting surface blemishes or even tearing. However, the adhesive-coated parts that are either in the mould cavities or sitting in a bin near the press should not be allowed to come in contact with the mould release that acts as a poison, thus hampering bonding. Part bins should be covered in case air currents carry the mould release onto the parts.

After vulcanisation, the bonded parts are usually submitted to additional treatments that could cause some common bond failures. For example, the cryogenic deflashing can induce failures in metal–rubber bond when large loads are in the tumbler at too low temperature for an extended period of time. Wire brushing, grinding or machining could determine the failure of the bonded part due to the heat build-up or to the mechanical action.

General references

[1]   Lindsay, JA. Practical guide to rubber injection moulding. Smithers Rapra, Shawbury, UK, 2012.
[2]   Packham, D. Handbook of adhesion. Wiley, Chichester, UK, 1989.
[3]   Crowther, B. Handbook of rubber bonding. Smithers Rapra, Shawbury, UK, 2001.

8 Manufacturing and transformation technologies

8.1 Designing rubber components

When a designer specifies the use of rubber for an object or a component, it is because no other material can duplicate the required performance characteristics. A wide understanding about the processes, materials and technical considerations involved in the design and manufacture of custom-moulded rubber parts allows a better control of costs. Part design should cover a wide range of parameters such as the required end performances, the quality, the conditions under which the item needs to operate and the standards it need to adhere to. For what concerns the required function of the part, for example, it could be useful to know if it has to seal a fluid and/or to be impermeable to particular fluid, or to transmit and/or absorb energy, or again to provide structural support. Regarding the environment, the part could operate in water, chemicals or solvents that could cause shrinkage, or in the presence of oxygen, ozone and/or sunlight, in wet/dry situation, under constant pressure or pressure cycle or under a dynamic stress, causing potential deformation. How long the part must perform correctly and what properties it must exhibit are additional key information. Some examples are the resistance to stretch without breaking (high ultimate elongation), to deformation (high modulus) and to set under extensive load (high compression set).

The selection of the most appropriate rubber material is a crucial factor in satisfying the design requirements for a given product. A wide range of basic elastomers are available and within these there are almost an infinite number of variations of formulations that can be produced. Therefore, selecting the best material means balancing suitability for the application, performance and ease of manufacturing. The rubber choice is also a key factor in determining the cost of the finished part. Specialised perfluoroelastomers, for example, can cost over a thousand times more than a basic natural elastomer. The manufacturing process to be employed is another factor affecting costs. Looking at the moulding technique, compression generally has the lowest mould costs, whereas injection the highest.

Cost needs to be balanced against performance requirements, taking into account the total cost of ownership. As well as the cost of raw materials and manufacture, the total cost of ownership estimation should include life expectancy, maintenance costs and the possible massive costs of component failure. A small reduction in price can lead to a disproportionate deterioration in quality.

https://doi.org/10.1515/9783110640328-008

Similarly, the tool life expectancy has to be considered in designing with rubber. It depends on many factors, including the mould material, the complexity of the mould, the required tolerance and the quantity of parts to be produced.

8.1.1 Tools helping to design with rubber

Designing elastomeric components is used to be referred to as a black art, due to the difficulties in assessing how the part would behave in service because of the unpredictable nature of elastomers. Thanks to the modern tools now at the disposal of engineers and to the progresses in the machining capability and manufacturing techniques, the design of engineering solutions using elastomers is now more than a predictable science. In this section, the role of these tools and some considerations required to ensure the right design are summarised.

Advanced product quality planning

Advanced product quality planning (APQP) is a structured method for defining and establishing the steps necessary to develop products that meet the customer requirements. It originates in the automotive industry, but it can be equally applied to product design across all sectors. APQP focuses on the up-front quality planning, and subsequently in the assessment if customers are satisfied, by evaluating the output and by supporting continual improvements. The approach consists of five main steps:

1. Planning and definition of the project
2. Design and development of the product
3. Design and development of the process
4. Product and process validation
5. Production launch; assessment and feedback about the process; eventual corrective actions

The final aim is to ensure an effective project design and management, in the view of the production readiness and set-up.

Finite element analysis

Finite element analysis (FEA) simulates the behaviour of a real component with an idealised mathematical model that includes the physical conditions in which it operates. The finite element model is then analysed by a solver, which calculates data reflecting the design behaviour to the applied boundary conditions, and helping to identify weaknesses or potential failures in the design. This technique is commonly used at the design stage for components but can also be used to help in the assessment of which and why parts have failed.

FEA of elastomer components is more complex when compared to thermoplastic or metallic parts, since both the chemical composition and the structure of elastomers make their behaviour more difficult to model.

CAD/CAM

The use of CAD/CAM systems for the design and manufacture of both parts and tools for rubber production may increase the productivity and can help to improve the quality of the design. It is particularly useful in the design of tooling for complex 3D parts. In addition to enabling drafting from sketches on existing components, CAD/CAM systems accept design data in a variety of proprietary file formats. Data can be supplied in the form of 2D drawings or as 3D models. Two-dimensional drawings are built up into 3D models by extruding, revolving or sweeping the 2D representation in space to create the base solid feature of the design.

When built up, the model can be visualised in three dimensions under simulated operating conditions and with the addition of any component. Parametric 3D modelling uses parameters to define the model features, such as its length or radius, and the geometric relationships among the constituent parts such as the relative position and tangency. This makes easy to modify the model keeping into account the elastomer shrinkage, for example, while ensuring that the required relationships remain as specified.

The production of complex 3D shapes on machining centres is made easier by the use of five-axis milling machines. By enabling the workpiece to be rotated, the required complexity and accuracy of the part can be produced in one setting, avoiding the need to use spark erosion in many cases.

Before moving to full production of tooling on a machining centre, CAD/CAM can be used also to produce a rapid prototype of the component for assessment purposes (visual inspections and specific tests). The prototype, which can be made in a variety of materials, mimics the function of the part. Prototyping is particularly appropriate with elastomeric materials since their complicated chemical nature is generally harder to predict their performance than for plastics or metal components.

8.1.2 Rubber over-moulding

Steel, brass, aluminium or plastic subcomponents are commonly incorporated into over-moulded rubber parts. These subcomponents are commonly termed inserts, as they are inserted into the mould. Typical metal inserts include screw machine parts, metal stampings and powdered metal shapes. Careful design of the insert can help to ensure a durable finished part while minimising the production costs.

Rubber can be over-moulded onto the insert by means of mechanical or chemical bonding. Mechanical bonding involves the incorporation of holes, depressions or projections in the insert itself. The rubber flows around or through the insert during the moulding process to create a mechanical assembly. Generally, when designing rubber over-moulded parts, the surface of the insert should be encapsulated in rubber as much as possible, with a minimum specified rubber thickness of 0.5 mm. This coverage helps to ensure the maximum bonding and controls the flash formation.

Special adhesives can be applied to the insert prior to moulding to create a strong chemical bond, as described in Chapter 7. In this case, the production of moulded rubber parts containing inserts typically involves considerable preparation before and after moulding. Steps may include cleaning and etching of the insert surfaces, masking and unmasking, application of adhesives and deflashing.

8.2 Rubber selection

In the rubber selection for specific applications, a number of criteria need to be considered, including the expected service conditions, the chemical compatibility with service fluids, the mechanical performances required during the service life, the suitability of the material for the application in a given temperature range, the exposure to temperature, ultraviolet (UV) light and sunlight, the contact with the human body (directly or indirectly, and for how long a period), the life prediction and some design considerations. Choosing the correct material always involves trade-offs in performance, as illustrated in Tab. 8.1. The key then is to determine and prioritise the most critical characteristics in order to do the best choice for a given part.

Tab. 8.1: Correlation of the main vulcanisate characteristics.

An improvement in...	Usually improves...	But sacrifices...
Abrasion resistance	Hardness/elongation	Resilience
Impact resistance	Elongation	Modulus
Creep resistance	Resilience	Flex resistance
Oil resistance	Tear resistance	Low-temperature flex
Resilience	Creep resistance	Tear resistance
Tensile strength	Modulus	Elongation
Vibration damping	Impact resistance	Structural integrity

Elastomeric materials are available in a wide variety of hardnesses, from 20 to 90 Shore A for thermoset rubbers, to even harder materials (Shore D scale) for thermoplastic elastomers. The selected hardness depends upon the final application. The most common hardness ranges from 50 to 80 Shore. Parts with complex geometry or deep undercuts are difficult to manufacture with very soft (<40 Shore A) or very hard (>80 Shore A) materials.

When choosing a rubber for a given application, it is important to understand whether it will be subjected to static or dynamic conditions. For example, if the material is subjected to dynamic forces then it may require enhanced abrasion resistance and excellent thermal conductivity properties. The application details required for a dynamic application would include if it is a rotary, reciprocating or vibrating environment. It would also be important to understand if the application would be subjected to thermal cycling, as this small amount of dynamic movement may also need to be considered in selecting the correct elastomer.

When selecting the appropriate rubber for applications involving temperature, it is essential to know the maximum and minimum continuous operating temperatures, the intermittent maximum and minimum exposure to temperatures and times, and the system pressure for low-temperature applications, if there will be thermal cycling, and the environmental factors to which the elastomer will be exposed. The upper temperature at which an elastomer can be used is generally determined by its chemical stability. Elastomers can be attacked by oxygen or other chemical species, which increase their effects with temperature. When elastomers are cooled to sufficiently low temperatures, they exhibit the characteristics of a glass, including hardness, stiffness and brittleness, and do not behave in the readily deformable manner usually associated with elastomers, loosing flexibility. As temperatures are raised, the segments of the polymer chain gain sufficient energy to rotate and vibrate. At high enough temperatures, full segmental rotation is possible and the material behaves in the characteristic rubbery way. The usefulness of an elastomer at low temperatures is based on whether the material is above its glass transition temperature, where it will still behave elastically, or below its T_g, where it will be hard and relatively brittle.

Fluids can affect elastomers both by physical and chemical interactions. The degree and type of physical interactions depend on a number of factors, including cross-link density and type, filler amount and type, polymer nature, type and viscosity of the fluids and solubility parameters of both polymer and fluid. The effects of physical interactions of fluids are normally observed as the rubber swelling. This process is generally reversible. The magnitude of swelling depends on the fluid, the elastomer and the temperature, and reflects the readiness with which the elastomer and its surroundings mix (i.e. the relative magnitudes of the solubility parameters of the two components). If a fluid has a solubility parameter close to that of an elastomer, the attraction and the mixing potential will be high, and a high volume swell results. The level of volume swell decreases as the difference in solubility parameters between the elastomer and the fluid increases. The high volume swelling determines the degradation of some physical properties, such as tensile strength, modulus and tear strength, and the reduction of the hardness.

Many chemical species cause degradation to the rubber compounds, such as acids, bases, water, hydrogen sulphide, zinc bromide, oxygen, ozone, mercaptans, free radicals and biocides as well as UV and ionising radiation. Degradation caused, for example, by water and amines is irreversible. It often results in rubber hardening or softening, increased compression set, cracking, and in extreme cases, the complete destruction. Such degradation is often highly dependent on the exposure to temperature on the rate at which the reaction proceeds.

8.3 Rubber moulding considerations

Manufacturing of rubber parts is usually accomplished in one of three ways of moulding: transfer moulding, compression moulding or injection moulding and their variations (Tab. 8.2). The choice of the most suitable technique depends on the

Tab. 8.2: Characteristics and application of moulding techniques.

Technique	Characteristics	Applications
Compression moulding	Pressing pre-shaped rubber compound in the mould + Tool costs − Manufacturing costs	Small series Large parts Prototypes Samples
Injection-compression moulding	Pressing pre-plastified rubber compound + Compound division without canals − Tool know-how	Large surface membranes and sealings Small precision parts High number tool parts
Transfer moulding	Transfer of rubber compound to cavities + Tolerances − Compound waste	Complex precision parts No secondary finishing (less flashes)
Injection moulding	Injection of pre-plastified material + Manufacturing costs − Tool costs	Medium-to-large series of parts Rubber-metal parts Automated manufacturing
Transfer-injection moulding	Transfer of pre-plastified rubber compound + Constant flow stream − Tool costs	Large series of small parts, also in rubber-metal. No secondary finishing (less flashes)
Multicomponent injection moulding	Injection of multiple materials during a cycle + Manufacturing costs large series − Tool costs, investment costs	Very large series Automated production No secondary finishing

number of factors, including size, shape and function of the part, quantity of items to be produced, type and cost of the raw material. The three methods, however, share certain basic characteristics that are important to understand when designing custom-moulded rubber parts. The goal is to select the mould design and the process that most closely approximates the optimal production conditions and cost requirements. The more demanding the part design, the more critical it becomes the mould making. The upfront investment in a costlier mould may pay for itself very quickly through lower material costs or more improved handling procedures.

After the product design, moulds are drawn and built by a mould-maker (or toolmaker) using precision-machines to form the features of the desired part. The recommended mould configuration depends on the size and complexity of the part, the hypothetic production volumes, the type(s) of material involved and the part function. Since moulds are expensive to manufacture, they are usually used in mass production where thousands of parts must be produced.

Typical moulds are built from hardened steel, pre-hardened steel, aluminium and/or beryllium–copper alloy. The choice of the material to build a mould is primarily based on economic considerations. Steel moulds cost more, but their longer lifespan will offset the higher initial cost over a higher number of parts made before wearing out. Pre-hardened steel moulds are less wear-resistant and are used for lower volume requirements or larger components. Hardened steel moulds are heat treated after machining; these are by far superior in terms of wear resistance and lifespan. Aluminium moulds are cost-effective in low-volume applications, as mould fabrication costs and time are considerably reduced. Beryllium–copper alloy is used in areas of the mould that require fast heat removal or in areas where the most shear heat is generated.

A mould consists of two or more plates, where the rubber compound is placed or injected between them. Plates are carefully registered to ensure consistent close tolerances and appropriate surface finish. These plates are exposed to heat and pressure to cure the part. The right time, temperature and pressure depends on the moulding process and the material.

Moulds can have a single cavity or multiple cavities. The real advantage of a single-cavity mould is that it lets change part design or material at minimal cost before committing to production. In multiple cavity moulds, each cavity can be identical and form the same parts or can be unique and form multiple different geometries during a single cycle. In designing multi-cavity moulds, it is important to quickly load the mould to avoid scorching. If the total loading time is too long, the moulding temperature may have to be reduced, and this will lengthen the moulding cycle, reducing the advantage of using multiple cavities.

Moulded products shrink during vulcanisation and oven post-curing. Since the coefficient of thermal expansion of rubber is greater than that of the mould material, when the moulded part cools to room temperature, it is smaller than the mould cavity at room temperature. Again, during vulcanisation and oven post-curing, decomposition products coming from the vulcanising agent and volatile components driven off from the rubber, thus decreasing the size of the moulded items. As a consequence, vulcanisates are smaller than the mould cavities in which they are formed. Since the dimensional accuracy of moulded rubber parts is very important, the mould design must allow for the shrinkage of the parts. The amount of linear shrinkage ranges from about 2% to 5%, depending on the rubber, the moulding temperature and the size and shape of the part. For example, silicone rubber shrinks more than organic rubber during vulcanisation.

With parts of complex shape, shrinkage is difficult to predict, since it depends on several factors acting at the same time, such as

- tool temperature and demoulding temperature;
- pressure in the cavity;
- location of the injection point (shrinkage in the direction of the material flow is usually somewhat higher than that perpendicular to the direction of flow);
- size of the part (the shrinkage of thicker articles is smaller than that of thinner articles);
- post-curing the article causes additional shrinkage of about 0.5–0.7%.

As shrinkage in elastomers is a volume effect, complex shapes in the moulded product or the presence of inserts may have the effect of restricting the shrinkage in one dimension and increasing it in another. However, when designing inserts for moulding to elastomers, other factors need to be considered, such as the fit in the mould cavities, the location of the inserts with respect to other dimensions, the proper hole spacing to match with mould pins and the fact that inserts at room temperature must fit into a heated mould.

8.4 Moulding problems

The proper troubleshooting should use a systematic approach to resolve problems in any moulding process. There are two kinds of issues: those involving the quality control and those encountered during start-up. Quality control issues occur when parts have been successfully produced in the past, but are now out of specification. These problems are the result of something in the process changing, such as material, machine and/or tool maintenance.

Start-up problems occur during the launch of a new tool or machine. To resolve them, the material process window must first be determined to ensure that there is a set of conditions that can make good parts. The starting point should be the setting of the process conditions to the middle of the material process range that are then adjusted to fix any observed problems. Sometimes a combination of variables must be changed to resolve the problems, such as material selection, machine selection and/or tool re-design.

8.4.1 Scorch

Scorch is the premature vulcanisation of the rubber before its flow in the mould is completed. The poor mould flow results in distorted or incompletely formed parts, showing heavy flash and incomplete cavity filling, cured ripples or texture at the surface and knit lines.

Scorch may be caused by hot spots in the mould, by moulding at too high a temperature, by too great a length of time between the starting of mould loading and the mould closing, or by restrictions in the mould that substantially reduce the flow rate. To prevent or eliminate scorch, the mould has to be completely filled before the rubber starts to vulcanise. More specifically:

- The mould cavity must be uniformly heated.
- If possible, a vulcanising agent with a higher reaction temperature must be used. This will bring the rubber to a higher temperature, and therefore to a lower viscosity for having a better mould flow.
- In general, with simple shapes, high mould temperatures and short cycles can be used; but with intricate shapes or thin items, lower temperature may be necessary. In compression moulding, the scorch is reduced keeping the moulding temperature near the low end of the vulcanising temperature and/or increasing the moulding pressure to increase the flow rate.

8.4.2 Backrinding

Backrinding corresponds to the distortion of the moulded product at the mould parting line, usually in the form of a torn or a ragged indentation. Typically, the flash in this area is heavy. Backrinding is most commonly experienced in compression moulding, but can happen in transfer and in injection moulding processes as well.

Backrinding may be caused by burrs or roughness of the mould parting edges, by a warped mould that does not close completely, or by moulding at too high a temperature for the vulcanising agent used. In the latter case, backrinding is caused by a sudden release of internal pressure within the part when the mould is opened.

To prevent backrinding, there is no roughness or excessive opening at the mould parting line. In addition, the moulding temperature has to be kept as low as possible depending on the vulcanising agent used.

8.4.3 Entrapped air

Air entrapped in the mould or in the rubber may produce soft, discoloured areas on the surface or in the cross sections of the moulded part, due to incomplete vulcanisation. Therefore, moulds should include provision for the release of air trapped in the mould cavity, by designing them so that they split in undercuts or sharp corners.

The clearance between mould parts should be large enough to allow the air to escape from the mould cavity, but not so large that the rubber also flows out. Generally, smooth-machined surfaces are satisfactory. The air that is enclosed in the cavity is first compressed by the injected rubber and then expelled through the venting channels inserted into the parting line so that the air can escape.

Optimum venting is created by vacuum as well. To produce such a vacuum, the mould stops during the closing movement at 0.5–2 mm before it is completely closed. A gasket is built into the parting line, so a vacuum pump can draw the air from the cavities. Only when the vacuum has reached a certain reduced pressure, the machine closes completely the mould and the injection process is started.

8.5 Post-moulding operations

After moulding, a number of operations may be needed to finish the vulcanisate. These include post-curing, removal of flashes and injection sprues, trimming, testing, inspection and packaging. It is worth to note that scraps of vulcanised rubber are not recyclable.

Removal of the waste edge or flash from a moulded rubber product can be accomplished in a number of ways. Depending on the rubber, part size and shape, tolerance and quantity, the deflashing methods commonly include the manual tear trimming and the cryogenic finishing. Cryogenic deflashing is a particular trimming technique that removes the excess imperfections on moulded parts using liquid nitrogen, high-speed rotation and some inert media (shot blast) in varying combinations to remove the flash in a highly precise and expedient manner. The part is cooled below its glass transition temperature, tumbled and blasted with beads to remove the flash.

Post-curing is one of the principal tools to mitigate outgassing and to remove the volatiles from the cross-linked rubber by diffusion and evaporation. Post-curing also helps to improve the compression set and generally the rubber mechanical performances giving the best dimensional stability. For many heat-resistant elastomers, such as fluorocarbons, HNBR and silicone, the oven post-curing is necessary to eliminate residues from peroxides and to complete the curing process.

With some materials post-curing in an autoclave is required, since it provides a comprehensive 'wash' of the vulcanisates especially for food or pharmaceutical applications. Another use of autoclaves is to cure products (not to be confused with post-curing) that are too big or unsuitable to be moulded, such as extruded parts and sheets.

In some cases, the manufactured products need to be non-destructively tested to ensure they meet the required specifications. Destructive testing of representative samples is also often carried out as an alternative to test the final parts.

One of the final steps in the manufacturing process is the inspection of finished part for quality control. This can be carried out by hand, with an inspector visually examining and measuring the vulcanisates. Alternatively, for reasonably simple components, inspection can be performed by machine, using a contact or non-contact system, working fully automatically, especially in the case of high-volume production runs such as o-rings.

The finished products need to be packaged appropriately before shipping. Some components may require to be sealed against moisture or contamination by other fluids, or protected against UV light.

8.6 Injection moulding

Based on the process intended for plastics moulding, injection moulding of rubber began in the mid-1960s. In rubber injection moulding, the material is heated and placed under a significative pressure per square inch of the cavity surface. This process is different from that of plastics, where the polymer is cooled in the mould under less pressure.

Parts to be injection moulded must be very carefully designed to facilitate the moulding process. The used material, the desired shape and features of the part, the material of the mould and the properties of the moulding machine must all be taken into account. The versatility of injection moulding is facilitated by this breadth of design considerations and possibilities. Commonly, injection moulding is ideal for high-volume productions of moulded rubber parts of relatively simple configuration.

An injection moulding machine is similar to an extruder. The main difference between the two machines is in screw operation. In the extruder type the screw rotates continuously providing output of continuous long product (pipe, rod and sheet). The screw of the injection moulding machine is a reciprocating screw since it not only rotates but also moves forward and backward according to the steps of the moulding cycle. It acts as a ram in the filling step when the compound is injected into the mould and then it retracts backward in the moulding step. The mould is equipped with a heating system providing a controlled heating during the vulcanisation process.

The injection moulding process occurs in six consecutive steps:

1. The uncured rubber compound is fed into the injection barrel in the form of a continuous strip.
2. The uncured rubber compound is pre-heated by the screw in the temperature-controlled barrel.
3. As the material accumulates in front of the screw, the screw is forced backwards. When the screw has moved back of a specified amount, the injection press is ready to shoot or inject the rubber compound, accurately metered, into the mould cavities and/or in the runner system.

4.   With the mould closed under hydraulic pressure, the screw is pushed forward. This forces the compound into the mould cavities and/or in the runner system for starting vulcanisation, by controlling the pressure, injection time and temperature.
5.   While the rubber cures in the heated mould, the screw turns again to refill the injection barrel.
6.   The mould opens and the vulcanised part can be removed. The injection machine is now ready to make its next shot as soon as the mould closes.

For any given rubber, the following production factors must be properly balanced in injection moulding for reaching the best results through experience and experimentation:
–   Barrel temperature to plasticise the rubber.
–   Moulding time, depending on the type of rubber and the size of the part.
–   Injection pressure, depending on the viscosity of the rubber, size of injection nozzle, mould design and desired injection time.
–   Injection time, depending on the mould cavity size, injection pressure and rubber viscosity. A short injection time, between 5 and 10 s, is desirable to minimise scorch and the total moulding time. Since it is important to keep injection time as short as possible higher pressures should be used with high viscosity rubbers.

Injection moulding machines can have a manual mode where the operator removes the moulded item at the end of each cycle. More common are semi-automatic or fully automatic machines where conveyor or pick and place robots remove the moulded product after the cycle completion.

The advantages of injection moulding are its suitability for moulding delicate parts, the shorter cycle times compared with compression moulding, the high accuracy and control of shape of the manufactured parts, the efficiency in mass production of large number of identical parts, the high levels of automation that can be introduced in the process and the lower levels of flashes. The main disadvantages include the costs of the tool, the longer change over times resulting from the more complex tooling and the waste of material in the runner system when a hot runner system is employed. Material waste also occurs when jobs are run sequentially with either differing materials or different colours, which requires extensive purging of the machine.

8.7 Compression moulding

Compression moulding is the most common moulding technique used in the rubber industry. It involves pressing uncured rubber between heated moulds so that the rubber compound first fills the mould cavity before curing.

The advantages of compression moulding are the lower cost of the moulds, the reduced overflow, the large sizes of vulcanisates and the relatively quick mould change with respect to injection moulding. Compression moulding can be a cost-effective solution in situations where the tooling already exists and the cross section of the part is very large and requires a long cure time. In addition, it is often chosen for medium hardness compounds, or applications requiring particularly expensive materials.

Compression moulds are generally loaded and unloaded manually and the elastomer (called *blank*) is often placed cold into the cavity so the cure times are longer. Some difficulties can arise from the positioning of the blank in the cavity and from the resulting flashes coming when too much material is placed in the cavity. Other disadvantages are the care and time required to manufacture the blank (weight and profile) to be placed into the cavity to have an efficient moulding run. Single cavity moulds are loaded by hand. With some multiple cavity moulds, loading boards may provide faster mould loading, which helps to prevent scorching of the pre-forms.

The compression moulding process involves the following steps:

1. A pre-weighed amount of the uncured compound is placed into the lower half of the mould.
2. The upper half of the mould moves downwards, pressing on the compound and forcing it to fill the mould cavity. The mould, equipped with a heating system, provides the compound curing by the combination of the hydraulic pressure and the temperature. Moulding time and temperature vary with the vulcanising agent used and the thickness of the part being moulded.
3. The mould is opened and the part is removed, ready for the necessary secondary operations, where flashes are trimmed.

8.8 Transfer moulding

Transfer moulding is a process in which a pre-weighed amount of a compound is pre-heated in a separate chamber (transfer pot) located between the top plate and the plunger and then it is forced into the pre-heated mould through a sprue, performing curing due to heat and pressure applied to the material.

Transfer moulding combines features of both compression moulding (hydraulic pressing) and injection moulding (ram-plunger and filling the mould through a sprue). As with compression moulding, transfer moulding requires secondary raw material preparation into pre-forms. It differs, however, in the placement of these pre-forms into the pot located between the top plate and the plunger. With respect to the conventional compression it also has simpler blank requirements and faster cure times, since the elastomer heats up quickly as it is transferred from the pot to the cavity.

The main disadvantages of transfer moulding are the higher cost of the tooling, the additional waste material due to the pot and the sprue and the difficulties that can be experienced when transferring high hardness or high-molecular-weight materials. The transfer moulding cycle time is shorter than that of compression moulding but longer that in injection. The method is able to produce more complicated shapes than compression moulding, but not as complicated as in injection moulding.

The transfer moulding process involves the following steps:

1. A piece of uncured rubber is placed into a portion of the mould called as transfer pot. The press plunger fits tightly into the pot, where the compound is heated and softens.
2. The plunger, mounted on the top plate, moves downwards, pressing the material and forcing it to fill the mould cavity through the sprue. In the mould, vulcanisation starts.
3. The press plunger is raised up and the transfer pad of the rubber compound is removed from the pot and thrown away.
4. The mould is opened and the part is removed from the cavity. Flashes and sprues are removed or trimmed.

Transfer moulding is particularly useful in producing parts whose shape is such that the moulds cannot provide good flow and tend to trap air. It is the best method of moulding parts that contain wires, pins and other ceramic or metallic inserts that require precise positioning in the mould cavity.

8.9 Rubber extrusion

Extrusion involves the continuous forcing of uncured rubber through a die of the required cross section under pressure, to form a shaped profile. The die is a sort of metal disk that has a machined opening in the shape of the part that needs to be extruded. The vulcanisation process takes place as the last step in the extrusion process. This aids the rubber-extruded profiles to maintain their shape and to acquire the required physical properties. Typical examples of extruded rubber parts are profiles, hoses, strips, sheets and cords. Rotating knives (or die face cutters) can then convert extruded materials into pellets or slugs for further processing.

Extruders are constituted by a screw housed within a barrel, with the screw turned by mechanical means. The uncured elastomer is first fed into the barrel via a hopper and then forced down the barrel by the screw while heat is supplied by the shearing action and via the heated barrel and screw. At the end of the barrel, in the extruder head, there is the die through which the material is forced out.

A typical phenomenon called die swell can occur when the rubber shape leaves the die. Because of this, the cross-sectional part becomes larger than the cross-sectional die. Thus, each die is made according to each particular part and material, to ensure that all tolerances are met for the finished extruded rubber part.

The extruded product can be vulcanised either in a heated pressure vessel (static vulcanisation) or by the continuous vulcanisation process. In the static vulcanisation the extrudate is conveyed from the extrusion machine to a station where it is cut into the required length and placed on a metal pan in a free state. So, it is not contained in a cavity as in moulding. The part is then vulcanised in an autoclave, heated by steam, for reaching the required temperature for the rubber vulcanisation. This is known as open steam vulcanisation. The pressure surrounding the extrudate during the open steam curing minimises the porosity formation.

In the continuous vulcanisation process the extrudate is fed directly from the extruder permitting the vulcanisation in a continuous length. Several media are employed in the continuous vulcanisation of rubber, all of which must be operated at elevated temperatures, like hot air, molten salts and microwave.

Microwave is a method whereby the extrudate is subjected to high-frequency electromagnetic waves that raise the temperature of the extrusion to near the curing state, uniformly throughout. The lack of pressure in most continuous vulcanisation processes makes difficult to control porosity.

Hot air vulcanisation is a continuous process in which the extrudate is passed through a horizontal chamber heated up to 500 °C or a vertical chamber heated up to 700 °C. Extrusions of small cross sections require only a few seconds to vulcanise at these temperatures, and those of larger cross sections require proportionately longer time. Both horizontal and vertical vulcanising chambers can be heated by strip heaters, infrared units, heating mantles or any other clean heat source. In addition, horizontal chambers can be heated with an airstream providing a very fast and efficient heat transfer. To save energy, part of the air stream can be conducted in a circulating system. All vulcanising chambers should have an air exhaust system to assure the proper removal of volatiles.

Horizontal units are the most widely used, and the best for producing extruded parts with one or more flat sides. Vertical units are the best for making thin-walled tubing, since there is no flattening in a vertical unit before that the vulcanisation is completed. In addition, vertical units provide a uniform heating around the extrudate, thus assuring a rapid and uniform vulcanisation.

A mixture of salts (KNO_3, $NaNO_3$ and $NaNO_2$) melted at 140 °C can be used as heat transfer fluid to vulcanise peroxide-cured compounds in the absence of oxygen. The extruded profile has to be kept under the surface of the molten salt by guiding rolls after leaving the salt bath. The salt solution has to be removed by washing the profile. The advantage of a salt bath for continuous vulcanisation is an excellent heat transfer and the lack of oxygen inhibition, even if the main hazards are fire and explosion associated with the use of molten salt at high temperature.

8.10 Calendering

A calender is similar to a mill and has two or more rollers (known as bowls) that can be adjusted to change the size of the nip which controls the thickness of the elastomer sheet. These bowls can be mounted horizontally or vertically and range in size from small laboratory devices to some weighing several tonnes. The material from the mixer is fed between the nips on the calender and pulled away from the bowl by a manual or mechanical device.

The calender process allows a high degree of control on the thickness of the rubber sheet that is achieved by adjusting the nips. Sheets are generally then used either to stamp a shape for placing into a mould in the next process, or to manufacture cross-linked elastomer sheeting from which gaskets or other finished products can be cut.

8.11 Tyre manufacture

Tyre manufacture is a very complex process since it involves the assembly of various rubber-based components, by placing them in the correct sequence and position around a driven rotating drum, to form an uncured or 'green' tyre carcass, which is then cured. The basic process is similar for all pneumatic tyres, from passenger and truck tyres to earthmover and aircraft tyres. A tyre is an assembly of many parts: a passenger car tyre has about 50 individual components, whereas a large tyre may have around 175 parts. Aircraft tyres are often hand built by two or more operators, whereas passenger and truck tyres are often mass produced on semi- or fully automatic machines, usually with a single operator. The rubber for tyres is compounded to provide traction, skid and cut resistance, and long wear.

Each tyre factory may use a known terminology to describe the various components of the tyre:

- *Inner liner:* The initial band of rubber strip forms the airtight inner impermeable surface of the tyre, in conjunction with the wheel rim. The inner liner acts like an envelope and retains the air pressure required to support the load.
- *Plies:* Strips of rubber reinforced with fine steel wires or synthetic (usually polyester, nylon or rayon) fibres, known as cords. Their main purpose is to provide the strength in the tyre. Car tyres tend to use fabric plies to provide the required flexibility, whereas in commercial tyres steel plies are more commonly used.
- *Beads:* Brass-plated, high tensile steel wire rings coated in rubber form the inner edge of the finished tyre. They anchor the tyre to the wheel rim.
- *Chafers (or apex strips):* Strips of rubber laid over the bead to cushion and protect it in service when in contact with the rim of the wheel.
- *Side walls:* Rubber strips placed over the chafers to form the side walls of the tyre, which provide good abrasion and environmental resistance for the sides of tyres.

- *Breakers (or belt plies):* Reinforced rubber strips, placed between the top plies and the tyre tread, give strength to the tyre allowing it to remain flexible in its structure.
- *Tread:* Unreinforced rubber strips improve wear and traction in any environment. In car and truck tyre production, tread is added as a single strip, but with earth-mover tyres it may be built up from several layers or extruded onto the carcass.
- *Carcass:* The uncured or 'green' tyre casing.

Tyre production can be summarised in three steps: (i) pre-forming of components; (ii) carcass building and addition of the rubber strips to form the sidewalls and treads; and (iii) moulding and curing of all the components into one integral piece. During the first a blank cylindrical casing without tread or breaker layers is produced. With the building drum in the collapsed position, the beads are placed by the operator onto the bead setting rings. The building drum is expanded and the inner liner placed on it, cut to length with a hot knife, and the joint stitched. The required number of plies is assembled on top, and the bead setting rings move in towards the drum to force the beads onto the carcass. After, strips of sidewall rubber are placed on the building drum, spliced and stitched together. This cylinder forms the first-stage carcass, also called first-stage green cover.

The second stage of the process may be carried out on the same tyre building machine (single-stage production) or the carcass can be transferred to another machine (two-stage production). The first-stage carcass is expanded pneumatically by the inflation bladder on the drum to form the final shape of the tyre. As the casing expands, two metal rings or bells move inwards to establish the final diameter of the casing. The breakers and then the tread are applied, spliced and stitched. The bells retract and the final casing is stitched, removed from the machine and moulded in a curing press. The tyre building machine pre-shapes radial tyres into a form very close to their final dimensions to make sure that the components are in their proper positions before the tyre goes into the mould.

The curing press is where tyres attain their final shape and tread pattern (third stage). Hot moulds like giant waffle irons shape and vulcanise the tyre. The moulds are then engraved with the tread pattern and the sidewall marking. Tyres are cured at over 100 °C for 12–25 min, depending on their size, in order to have the simultaneous vulcanisation of the different rubber compounds. As the press swings open, the tyres are popped from their moulds onto a long conveyor that carries them to the final finish and inspection.

Professional visual inspector and automated inspection machines detect the slightest defect on the final product. Inspection does not just stop at the surfaces. Some tyres are sampled from the production line and X-rayed to detect any hidden weaknesses or internal failures. In addition, quality control engineers regularly perform cut sections to study all parameters that can affect performance, ride or safety on a tyre.

General references

[1] Morton, M. Rubber technology. Van Nostrand Reinhold, New York, USA, 1987.

[2] Eirich, FR., Coran, AY. Science and technology of rubber. Academic Press, New York, USA, 1994.

[3] Long, H. Basic compounding and processing of rubber. Lancaster Press, Lancaster, USA, 1985.

[4] Gent, AN. How to design rubber components. Hanser Publishers, New York, USA, 1994.

[5] Babbit, RO. The Vanderbilt rubber handbook. Vanderbilt Company, Norwalk, USA, 1978.

[6] Gent, AN. Engineering with rubber. Hanser Publishers, New York, USA, 1992.

9 Rubber materials and products testing

9.1 Introduction

Properties of rubber are drastically different from the other engineering materials [1]. Consequently, testing procedures are often unique and specifically dedicated. Depending on the final application, the most appropriate tests can be selected and carried out to give the best results to predict the life time of a given item. In the following sections, important properties and the related test methods are explained in more details.

9.2 Viscosity

Viscosity is defined as the resistance of a fluid, such as rubber, to flow under stress [2]. Mathematically, it is given by the ratio between the shear stress and the shear rate. It is dependent on the temperature: at higher temperature, any material is less viscous.

In rubber industry, viscosity is commonly measured using a rotational viscometer called Mooney viscometer developed in 1930s. In this test method, two pre-cut rubber pieces with a volume of 25 m^3 are placed into a two-part compression cavity mould. When the dies close, a pressurised sealed cavity is formed, in which a special rotor is embedded in the rubber. Usually there is a pre-heating time after the dies are closed to allow rubber to approach the selected curing temperature. Then, the rotor turns at two revolutions per minute (2 rpm) for a specified time period. The instrument records the viscosity in Mooney units (MU), which are arbitrary units based on the torque values.

Generally, the measured viscosity of rubber under test decreases with the running time because of the thixotropic effects of the material itself. This decrease is slow depending on the type of rubber and the test temperature. Usually, the Mooney viscosity is measured with 1 min of preheating and a run time varying from 4 to 8 min. The final Mooney viscosity value is reported at the lowest value recorded in the last 30 s of the test.

9.3 Cure characteristics

In compound development, the composition of any formulation can be varied until the desired vulcanisation characteristics are achieved [3, 4]. One of the most used and easiest way to monitor the curing process is rheometry, which has been used for many years to study how a rubber cures over time.

https://doi.org/10.1515/9783110640328-009

The rheometer is a key equipment in the rubber industry, which helps to investigate the curing characteristics of the rubber compound and to monitor the processing behaviour as well as the physical properties of the vulcanising material. The cure curve obtained with a rheometer is a finger print of both vulcanisation and processing. Interesting parameters such as the scorch time, cross-link degree and reaction kinetics can also be obtained from this test.

More specifically, rheometry has manifold advantages in different fields of rubber world:

-   *Research and development*: The most tedious part in compounding is to develop a new compound. This activity involves some steps, such as defining required quality targets, designing preliminary compounds, selecting specific ingredients and determining their dosage, checking the cost, testing each compound and redesigning the formulation till the quality target is achieved. This process implies an enormous work that is time consuming and expensive and requires well-defined skills. With the help of rheometry, one can do all these exercises quickly, with minimum waste of materials and time.
-   *Quality control*: To produce good vulcanisates, it is of vital importance that rubber is of consistent quality. As the compound is mixed in batches, batch-to-batch variation can occur; thus, attention has to be paid in controlling the quality of each batch. If randomly selected batches are subjected to the rheometeric analysis, upper and lower control limits, range, mean and standard deviation, with reference to the rheological parameters, can be defined. Each batch on testing can be classified on pass/fail criteria depending upon the quality control limits for that kind of rubber.
-   *Process control*: Rheometry gives a true picture regarding the processing behaviour of the rubber compound in terms of viscosity, scorch time and optimum cure time. Consequently, compounds are stored, processed and used accordingly. Unused compounds left for longer periods will tend to spontaneously cure, rendering it unsuitable for the future use. Such mixes can be tested easily on the rheometer and any decision can be taken regarding their utility. In addition, the ability of a rheometer to detect minor changes in the composition of rubber compounds has made it a widely accepted production control test equipment.
-   *Effects of new ingredients*: Rheometry is useful to carry out studies on changes of any chemical in an existing compound, since the effect caused by the ingredient modification can be immediately observed on rheometeric curve.
-   *Optimization of ingredients amount*: The need to change the amount of any ingredient in the compound can be noticed immediately by the rheometric analysis and, accordingly, the variation can be done.

Two different kinds of rheometer are available for measuring the cure profile properties of any rubber: the oscillating disk rheometer (ODR) and the moving die rheometer (MDR).

9.3.1 Oscillating disc rheometer

The ODR is an efficient, simple and reliable testing equipment that describes precisely and quickly the curing and processing characteristics of rubber compounds [3]. It was introduced in 1963 and was considered immediately over the Mooney viscometer because ODR measures not only scorch but also the cure rate and the state of cure.

ODR works on a very simple principle and operates quite easily. The uncured rubber sample is confined in a die cavity located between two electrically heated platens. The cavity is formed by a fixed lower die and a moveable upper die (Fig. 9.1). The dies are kept closed during the test by a pneumatic pressure ram. The die temperature can be selected within the range 100–250 °C. A rotor (biconical disc) is embedded in the test specimen and oscillates sinusoidally with a small specified rotary amplitude through an arc of 1°, 3° or 5°. This action exerts a shear strain on the test sample.

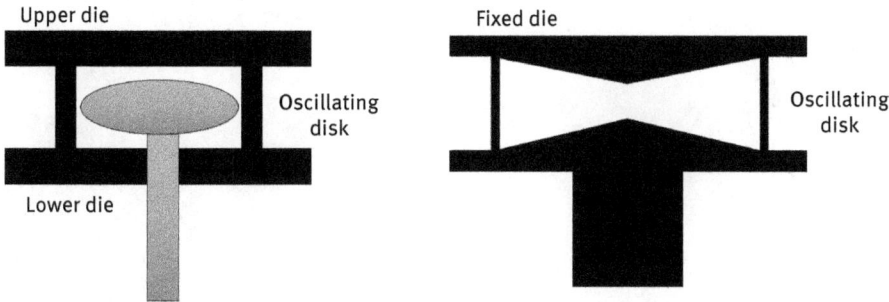

Fig. 9.1: Rheometer types: ODR (on the left) and MDR (on the right).

The force required to oscillate the disc depends upon the stiffness (shear modulus) of the rubber compound. Stiffness increases when the formation of cross-links during curing. The force required to oscillate the disc is measured electronically by the torque arm transducer, and it is continuously plotted on the recorder against time, thus resulting in the cure curve (torque vs time). The increase in the torque level during vulcanisation is proportional to the number of cross-links formed per unit volume of rubber.

The determination of the optimum cure temperature for a small specimen as in the rheometer is not the same as determining the optimal cure time for a rubber part cured in a factory. However, the rheometric analysis may supply a first insight to set up the key parameters on the moulding machine. Of course, in a factory other factors have to be kept into account, making experimental trials at different time and temperatures to achieve the best physical properties and performance in the final rubber items, considering also at the best productivity.

9.3.2 Cure curve

Fig. 9.2 shows a typical cure curve obtained with an ODR; from this we can directly determine all the vulcanisation characteristics of the rubber compound [3, 4]. During the analysis both in ODR and in MDR, the rheometric curve is displayed in real time and at the end of the test time results are automatically computed and displayed on the screen.

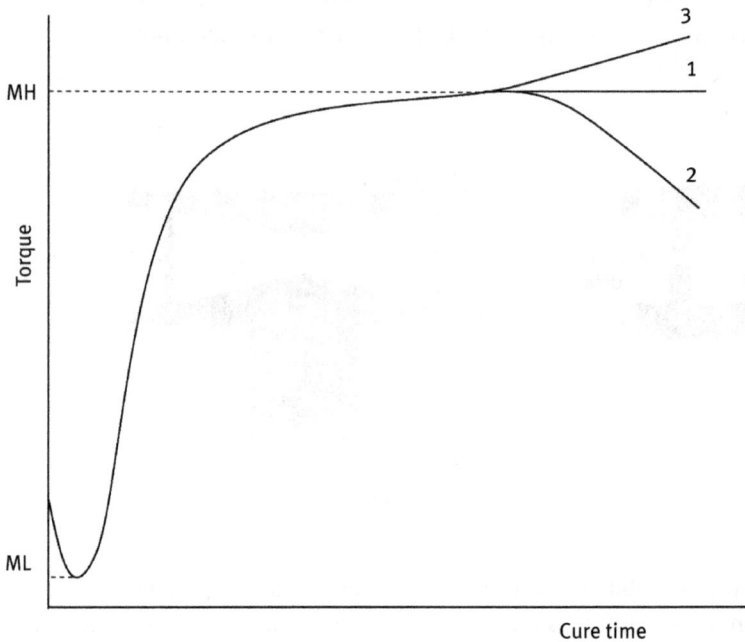

Fig. 9.2: Examples of cure curves: (1) NBR, (2) natural rubber and (3) HNBR.

The time required to obtain a cure curve is a function of the test temperature and the vulcanisation characteristics of the rubber compound. Thus, the cure curve represents the development of properties in a rubber compound at a given curing temperature. A very good correlation can be obtained for a compound or a family of compounds where the curing system is the only variable. No direct comparison can be made when the polymer and the level of filler loadings are different.

The rheometric curve can be divided into three phases:
- Phase 1: It gives an indication of the processing behaviour of the compound in terms of scorch time.
- Phase 2: It describes the curing characteristics of the compound (cross-linking).
- Phase 3: It gives a good idea about the physical properties of the vulcanised compound, since the cross-linked network reached its maturation.

Both ODR and MDR produce rheometric curves showing all the three phases with a characteristic shape. A trained eye can monitor the initial trough, the slope of rise during the curing phase and the final shape of curve.

From the rehometric curve, we can obtain some parameters that give information about the reaction kinetics and the cross-link density:

Torque values:
- MI (initial torque): It is the torque recorded at the start of the test and represents the initial viscosity of the system.
- ML (minimum torque): As the compound gets heated under pressure, the viscosity decreases and the torque falls to the lowest value, called ML. Basically, it is a measure of the stiffness and viscosity of the unvulcanised compound and it can be related to the dispersion of the fillers and their interaction with the polymer matrix.
- MH (maximum torque): As the curing starts, the torque increases proportionately. Depending upon the compound, the slope of rising torque varies. Torque typically reaches a maximum value (MH) and after to a plateau behaviour in which the physical property, after attaining the maximum value, remains constant with continued cure. If the test is continued for sufficient time, the reversion may occur and torque tends to fall. Generally, on prolonged cures the physical properties of vulcanisates will start deteriorating.
- MH-ML: The difference between the maximum and the minimum torque relates to the cross-link density of the material.

Time values
- ts2 (induction time): It represents the scorch time when viscosity rises 2 units above ML.
- ts5 (scorch time): It is the time for viscosity to rise 5 units above ML (5% increase of the initial torque). Both ts2 are ts5 are measures of the initial slope of curing phase that measure the scorch time and, consequently, the processing safety.
- tc50: It is the time at which 50% of cure has taken place.

- tc90 (optimum cure time): It is the time at which 90% of cure has taken place and at which 90% of the final torque is reached. It is considered the optimal time for the rubber vulcanisation as overheating the material could lead to reversion processes.

Derived values
- Cure rate: The cure rate is essentially a measure of the linear slope of the rising curve and refers to the amount of time required to reach a specified state of cure at a given vulcanisation temperature or heat history. It is the rate at which cross-linking and development of stiffness (modulus) of the compound occur after the scorch point. As the compound is heated beyond the scorch point, the properties of the compound changes from a soft plastic to a tough elastic material required for use. As more cross-links are introduced, the polymer chains become more firmly connected and both hardness and modulus increase.
- Reversion time: It is the time to reach 98% MH after passing MH. The reversion time gives an indication of the quality of the compound as to how long it retains its physical properties when subjected to heat ageing. Reversion occurs with over-cure, and both the modulus and tensile strength decrease. This phenomenon is observed in the vulcanisation of natural rubber, polyisoprene and butyl rubber. Other rubbers such as NBR, HNBR, styrene–butadiene rubber and neoprene generally do not show reversion.

Fig. 9.2 shows three different types of cure curves obtained with different types of rubber compounds. Curve 1 belongs to a synthetic rubber compound (i.e., NBR and silicone) that has attained a constant torque MH. Curve 2 is related to natural rubber that has attained the maximum torque and after which it reverts. Curve 3 is for HNBR that shows an increase in torque with further cure. In this case, the compound continues to harden, the modulus rises and the tensile strength as well as the elongation continue to drop.

9.3.3 Moving die rheometer

Even though thousands of ODRs have been used worldwide, this kind of rheometer has a design flaw that involves the use of the rotor; there are some problems associated with the same [3]:
- The ODR disk itself works as a 'heat sink', preventing the rubber specimen from reaching the curing temperature quickly. Since the rotor is not heated, the ODR scorch values are not the true isothermal measurements.
- The ODR torque signals must be measured through the shaft of the rotor. This design results in poor signal-to-noise response due to the friction associated

with the rotor itself. This friction prevents the measurements of the true dynamic properties during cure.

- Since rubber embeds the rotor, ODR dies are easily fouled and require constant cleaning. This also hampers the automation of tests because the cured specimen must be manually removed from the rotor at the end of each analysis.

These problems have been significantly reduced with the introduction of a new rotorless curometres: the MDR, in which the lower die oscillates sinusoidally, applying a strain to the rubber specimen that is constrained in a sealed pressurised cavity (Fig. 9.1). The upper die is attached to a reaction torque transducer, which measures the torque response.

Therefore, MDR is basically an improved ODR where several limitations of ODR have been overcome. The advantages of MDR over ODR include the reduction of time lag between the test sample temperature and the die cavity temperature and the friction between the rotor and the test cavity by the elimination of the rotor itself. This design greatly improves the test sensitivity (signal-to-noise) measurements and so the real changes in the rubber compound can be detected faster.

The cure times for MDR are significantly shorter than those measured with ODR for the same compound in the same curing conditions. This is because MDR induces a cure process that is closer to a true isothermal cure. The temperature recovery for MDR is only 30 s, compared to approximatively 4–5 min for the ODR, and the temperature drop when loading a sample is lesser than for an ODR. Therefore, cure kinetic studies can be performed much better with MDR.

Since the torque is measured in a reactive manner, both the elastic and viscous responses can be separated and measured simultaneously with the cure data throughout the test. Thus, two curves are recorded during cure (Fig. 9.3). The elastic torque S' is the traditional cure curve, which is commonly used as an indication of cure state (the ODR produces only an S' curve). The viscous torque S'' curve is a second trend generated simultaneously, which provides information about processing characteristics. In some cases, the S'' peak can be used as an alternate method for measuring scorch. The viscous response S''(out of phase with the strain applied) is typically 0.1–0.5 times of the elastic response S' (in phase with the strain applied). Tan delta (tanδ = S''/S'), measuring the dissipation factor, can also be in real time plotted on the MDR since both S' and S'' values are measured continuously.

A more advanced rotorless curometre is the rubber process analyser. It is designed for measuring the viscoelastic properties of polymers and elastomeric compounds before, during and after cure. The acquired data give exact information about the processability, cure characteristics, cure speed and the behaviour of the compound after cure.

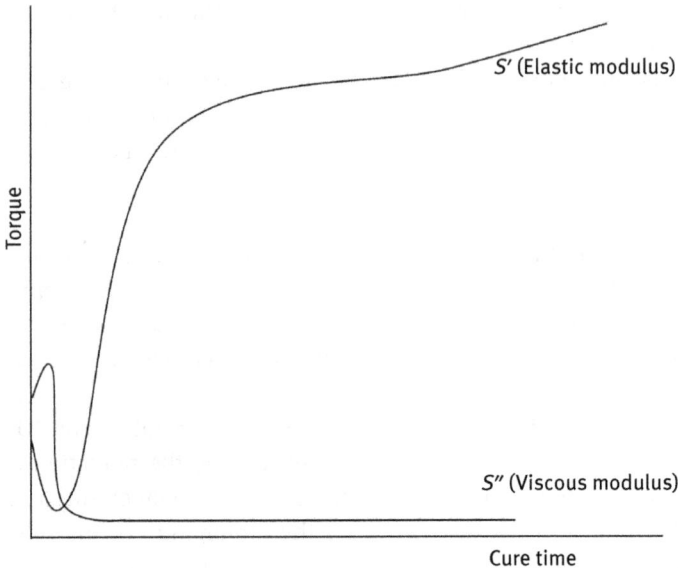

Fig. 9.3: Rheometric curves recorded by MDR.

9.4 Hardness

Hardness is defined as the resistance to indentation and represents the elasticity of the material [5]. The lower the hardness, the more elastic is the material. Hardness measurements in rubber are expressed in Shore A or Shore D units.

The instrument used for the measurement of Shore A hardness is the durometer. It measures the depth penetration of a stress-loaded metal sphere into the rubber. The harder the sample, the farther it can push back the indenter point and the higher will be the numerical reading on the scale. In the Shore D tester, a dead load is applied to the indenter for a specific time.

Because of the viscoelastic nature of rubber, a durometer reading shows the maximum value as soon as the metal sphere reaches maximum penetration into the specimen and then decreases in next 5–15 s. Hand-held spring-loaded durometers are commonly used but are subject to operator error. Bench-top dead-weight-loaded instruments reduce the error to a minimum.

9.5 Tensile strength and elongation

The stress–strain properties of rubber compounds are usually measured under tension following the ASTM standard D412 [6, 7]. Tensile strength is the maximum tensile stress reached during stretching a test specimen, usually a flat dumb-bell shape, up to its breaking point. By convention, the required force is expressed as force per unit area of the original cross-section of the test length. Tensile strength is a useful quality control tool to monitor the inter-batch consistency. It does not, however, give any indication of extrusion resistance.

Elongation, or strain, is the extension between bench marks in the test specimen produced by a tensile force applied to it. It is expressed as a percentage of the original distance between the marks. Elongation at break, or ultimate elongation, is the elongation at the moment of breaking.

The tensile behaviour of a rubber shows a completely different trend with respect to the metal for which the ratio between stress and strain is a linear relationship, as shown in Fig. 1.3. For rubber, at very low strains, the ratio of the resulting stress to the applied strain (Young's modulus) is a constant. This value is the same whether the strain is applied in tension or compression. Hooke's law is, therefore, valid within this proportionality limit. However, as the strain increases, the linearity ceases, and Hooke's law is no longer applicable.

Stress measurements are made at a specified percentage of elongation and are given as modulus values. For example, 300% modulus is defined as the stress per unit cross-sectional area (in psi or MPa units) at an elongation of 300%.

Temperature has a marked effect on the strength properties of elastomers (tensile, flexural or compressive). Room temperature testing rarely gives an accurate indication of their strength at elevated temperatures: for example, at 100 °C some elastomers retain only 10% of their room temperature strength. To obtain a more meaningful result, tensile testing can be performed at elevated temperatures in suitable testing machines.

9.6 Tear strength

Tear strength is a measure of the resistance of an elastomer to tearing [6]. It is measured using a tensile test machine operating at a constant rate until the test piece breaks following the ASTM standard D624 [8]. Various types of test pieces can be used and, depending on the method employed, the maximum or median force achieved is used to calculate the tear strength.

Tear-testing procedures concentrate the stress in one area, either through sample design or by cutting a nick in the sample. The peak force and sample thickness are recorded. Tear values are reported in units of kilonewtons per meter.

9.7 Compression set

Rubbers deform under load and rarely return completely to their original dimensions when the load is removed [9]. The difference between the original and the final dimensions is known as the compression set. Thus, the compression set can be described as the ability of an elastomer to recover from an imposed strain.

Standard methods for compression set measurements require a compressed sample to be exposed for a fixed time, at a given temperature, and then allowed to recover at room temperature. Small cylindrical disks of 13 mm diameter and a thickness of 6 mm, or 29 mm diameter and a thickness of 12.5 mm, are being used to perform the tests. The disks are compressed in such a way that the compression is equal to 25% of the original height at a known temperature (often at 23 °C) for a duration of 24 or 72 h. At the end of the specified time, the test pieces are removed from the test jig and allowed to recover at 23 °C for 30 min before the thickness is re-measured. The compression set is measured as the difference between the original thickness of the test piece and that after recovery, expressed as a percentage of the initially applied compression: 0% indicating full recovery and 100% indicating no recovery.

9.8 Abrasion test

The abrasion test is used to measure the resistance of a material to wear stemming from sliding contact such as rubbing, grinding or scraping against another material [9]. Abrasion tests try to accelerate the process by applying more cutting-like conditions. However, this approach may not simulate the real wear.

Abrasion may be measured in a number of ways, depending on the resistance test used and the information that is desired from the test. For example, where the amount of material lost is a concern regardless of whether the material fails, abrasion may be measured in terms of the percentage of material lost, either by mass or by volume, between the start and end of the run: a smaller number indicates a good abrasion resistance. Another test measures the number of abrasion cycles a material withstands before failure. This would be more appropriate if information on how long the material or product will survive before failure is of primary interest.

Several factors are typically considered in developing or selecting an appropriate abrasion test, such as the shape of the contact area and the composition of the two surfaces coming in contact with one another. Speed of sliding contact between the two surfaces, the force with which they act on one another and the duration of contact between them may also be considered. In addition to the materials themselves, the environment in which they are coming in contact also plays a role in selecting the most appropriate abrasion test.

It is sometimes wrongly believed that tensile strength is related to abrasion resistance, and while a high tensile strength compound can have good abrasion resistance, the converse can also be true. Abrasion resistance is related more to the polymer type and the nature/level of compounding ingredient used. High modulus and high tear strength can be better correlated to the abrasion resistance, but the relationships are not definitive.

9.9 Resistance to fluids

The effect of a fluid on a particular rubber depends on the solubility parameters of the two materials [4]. The more the similarity, the larger is the effect. A liquid may cause the rubber to swell due to absorption, it may extract chemicals from it or it may chemically react with it. Any of these effects can lead to a degradation of rubber. The effect of any liquid on rubber is determined by measuring changes in volume or mass, tensile strength, elongation and hardness after immersion in oils, fuels, service fluids, solvents or water.

Absorption is usually greater than extraction and it is linked to a net increase in volume, generally known as swelling. Since vulcanised rubber does not dissolve but swells in solvents, the extent of swelling depends on cross-link density. A tightly cured sample will swell less than a slightly cured sample. For some products, a decrease in volume or dimensions could be more serious than swelling, and if there is a significant chemical reaction, a low swelling may hide a large deterioration in physical properties. Consequently, although the degree of swelling provides a good general indication of resistance, it is also important to measure the change in other properties.

9.10 Low-temperature properties

All elastomers undergo several kinds of change when they are exposed to low temperatures [3, 9]. Some of the changes occur immediately, and others after prolonged exposure. At low temperatures, the material will become brittle and shatter on sudden bending or impact. The temperature at which this occurs, when determined under certain prescribed testing conditions, is called the brittle point.

The effect of temperature on rubber stiffness is measured using the Gehman apparatus. This test evaluates the torsion modulus by twisting a strip test piece, at room temperature and several reduced temperatures, to give a temperature–modulus curve. The result is often quoted as the temperatures at which the modulus is 2, 5, 10 or 50 times the value at room temperature. However, a more useful measure is the temperature at which the modulus increases to a pre-determined value, normally 70 MPa, which corresponds to the limit of technically useful flexibility.

Another test, to measure the modulus of the material, is the material retraction test. This test is carried out by elongating a test specimen and freezing it in the elongated position. The specimen is then allowed to retract freely, whilst the temperature is slowly raised at a uniform rate. The percentage retraction can be calculated at any temperature from the obtained data. In practice, the temperature corresponding to 30% retraction (TR30) roughly correlates to the limit of useful flexibility.

9.11 Resistance to weathering

Deterioration in physical properties can occur when rubbers are exposed to weather conditions [9]. This includes cracking, peeling, chalking, colour changes and other surface defects that ultimately may lead to failure. The most important causes of degradation are ozone, sunlight (UV), oxygen, moisture and temperature.

Environmental testing is used to predict how any material will behave in the normal conditions of use. Samples are placed in climatic chambers in a controlled environment, such as low or high temperatures, relative humidity, light or pressures for a determined time. Afterwards, they are removed, are allowed to cool and then tested and compared against the original properties for the material measured at the room temperature.

References

[1]   Morton, M. Rubber technology. Van Nostrand Reinhold, New York, USA, 1987.
[2]   Long, H. Basic compounding and processing of rubber. Lancaster Press, Lancaster, USA, 1985.
[3]   Dick, JS. Rubber technology-Compounding and testing for performance. Hanser, Munich, Germany, 2001.
[4]   Gent, AN. Engineering with rubber. Hanser Publishers, New York, USA, 1992.
[5]   Eirich, FR., Coran, AY. Science and technology of rubber. Academic Press, New York, USA, 1994.
[6]   Bartenev, GM., Zuyev, YS. Strength and failure of viscoelastic materials. Pergamon Press, New York, USA, 1968.
[7]   ASTM D412, Standard Test Methods for Vulcanized Rubber and Thermoplastic Elastomers.
[8]   ASTM D624, Standard Test Method for Tear Strength of Conventional Vulcanized Rubber and Thermoplastic Elastomers.
[9]   Brown, RP. Physical testing of rubber. Elsevier Applied Science Publishers, New York, USA, 1986.

10 Recycling rubber

10.1 Why reclaim or recycle rubber?

With the increase in demands, the manufacturing and use of rubber and rubber items have increased tremendously both in the developed and less developed countries [1, 2]. Therefore, one of the various problems that mankind faces nowadays is the problem of rubber waste disposal management.

Since rubber compounds do not decompose, their disposal is a serious environmental problem. About 242 million tyres are discarded every year in the United States alone. Less than 7% are recycled, 11% are incinerated for their fuel value and another 5% are exported. The remaining 78% are either landfilled, or are illegally dumped. Land filling with waste tires is the most unwanted due to related environmental problems and has no future possibility. Indeed, after discarding the tires for landfilling there is a probability of leaching small molecular weight additives from bulk to the surface and from the surface to the environment. These small molecular weight additives are not eco-friendly and may kill advantageous bacteria of soil.

Even if rubber recovery can be a difficult and complicated process, there are many reasons why rubber should be reclaimed or recovered:

- Recovered rubber can cost half than of natural or synthetic rubbers and may have some properties better than those of virgin rubber.
- Producing rubber from reclaim requires less energy in the total production process than does virgin material.
- It is an excellent way to dispose the unwanted rubber items.
- Recycling activities can generate work in developing countries.
- Many useful products are derived from reused tyres and other rubber products.
- If rubber goods are incinerated to reclaim embodied energy, then they can yield substantial quantities of useful power. For example, in Australia, some cement factories use waste tyres as a fuel source.

There are many ways in which tyres and other rubber products can be reused or reclaimed. The waste management hierarchy shown in Tab. 10.1 dictates that reuse, recycling and energy recovery, in this order, are superior to disposal and waste management options.

Damaged tyres are, more often than not, repaired. Tubes can be patched and tyres can be repaired by one of a number of methods. Regrooving is a practice carried out in many developing countries where regulations are slacker and standards are lower. It is often performed by hand and is labour intensive.

Secondary reuse of whole tyres is the next step in the waste management hierarchy. Some examples of secondary use in industrialised countries include applications

https://doi.org/10.1515/9783110640328-010

Tab. 10.1: Principal rubber recycling processing paths.

Kind of recovery	Process	Recovery method
Product reuse	Repair	Retreating
		Regrooving
	Physical reuse	Use as weight
		Use of form
		Use of properties
		Use of volume
Material reuse	Physical	Tearing apart
		Cutting
		Processing to crumb
	Chemical	Reclaim
	Thermal	Pyrolysis
		combustion
Energy reuse		Incineration

as tree guards, in artificial reefs, fences or as garden decoration. In developing countries walls can be lined with old tyres. Docks are often lined with old tyres that act as shock absorbers, and similarly crash barriers can be built from old tyres. Old inner tubes also have many uses, as swimming aids and water containers being two simple examples.

The next step in the waste management hierarchy involves the material being broken down and reused for the production of a new item. In developing countries, this hand reprocessing of rubber products to produce consumer goods is well established and the variety of products being made from reclaimed tyres and tubes is astonishing. Shoes, sandals, buckets, motor vehicle parts, doormats, water containers, pots, plant pots dustbins and bicycles pedals are among the manufactured products.

The two principal methods to obtain a re-usable recycled rubber material are: (i) grinding of the rubber and reusing it in the form of a granulate or surface activated powder and (ii) treating the material in a reclaiming process to generate a viscous-elastic reclaim. Granulate tends to be used for low-grade products such as automobile floor mats, shoe soles, rubber wheels for carts and barrows and so on, and to be added to asphalt for road construction. Different processes have been developed in order to reclaim vulcanised rubber. Chemical and thermal recovery is a higher technology requiring sophisticated equipment in which the waste rubber reclaims are treated with chemicals or by heat and then processed mechanically.

Sometimes scrap rubber is used as a fuel. Pyrolysis is one of the thermal approaches to recover energy from waste rubber. An environmentally friendly process was developed for recycling rubber waste materials, such as tires, to generate valuable fuels or chemical feedstocks in a closed oxidation process that is free of hazardous emissions [1]. The process involves the breakdown of rubber materials by selective oxidation decoupling of C–C, C–S and S–S bonds by water as a solvent at or near its supercritical condition.

10.2 Rubber reclaim

Reclaimed rubber is the product resulting when waste vulcanised scrap rubber is treated by heat, chemicals and mechanical techniques to produce a plastic material which can be easily processed, compounded and vulcanised with or without the addition of either natural or synthetic rubbers [3]. Regeneration can occur either by breaking the existing cross-links in the vulcanised polymer or by promoting the scission of the main chains of the polymer or a combination of both processes. During reclaim the molecular weight is reduced, so reclaimed compounds have poorer physical properties when compared to new rubber.

Although reclaim rubber is a product of discarded rubber goods, it has gained much importance as additive in various rubber formulations. It is true that mechanical properties like tensile strength, modulus, resilience, tear resistances and so on are all reduced with increasing the amount of reclaim rubber in fresh rubber formulation, but at the same time reclaim rubber provides many advantages if incorporated in fresh rubber.

In many products 5–10% reclaim can be added to the new rubber content without serious effects onto the physical properties. Higher percentages (20–40%) are used in products like car mats. Traditionally, compounds used in the production of tyre carcasses have been the main outlet for reclaim due to its processing advantages. In spite of this, the proportion of reclaim in radial tyres is limited to around 2–5%.

During the reclaiming process, rubber has already been plasticised due to a large amount of mechanical working. Therefore, in the consumer hands it mixes more easily than the new rubber in lesser mixing time with less heat generation. This is particularly advantageous with compounds containing high carbon black loading. Reclaiming of scrap rubber is, therefore, the most desirable approach to solve the scrap/waste rubber disposal problem, saving some precious petroleum resources and power.

In the mixing of tire carcass and side wall stocks, reclaim rubber is not added during the first Banbury pass, rather it is added during the second Banbury pass along with the curing agents. The second pass is much shorter than the first, therefore, an increase in mixing capacity of as much as 40% occurs with a 30% Banbury cost saving per kg of rubber.

Reclaimed rubber stocks can usually be processed at a lower temperature than those containing virgin rubber alone. It generally provides faster processing during extruding and calendering. Due to the presence of cross-linked gel, reclaimed rubber is less thermoplastic than new rubber compounds. Thus, when extruded and cured in open steam it tends to hold the shape better. Extruder die swell and calender shrinkage can be reduced with a proper use of reclaim rubber due to its lower nerve. Using reclaim rubber in tire carcass stocks permits high-speed calendering and results in smooth, uniform coating as well.

Reclaim rubber containing compounds help to retard and reduce the sulphur bloom from both uncured and cured stocks. It cures faster than virgin rubber compound, probably due to the combined action of sulphur and the active cross-linking sites already existing. Energy savings thus obtained constitute its usefulness in commercial purposes.

Rubber may be converted into reclaim by means of a number of processes: mechanical, thermal, thermo-mechanical and pyroltic [4]. In all these, the scrap rubber must first be shredded and ground into crumbs to permit the chemicals and swelling agents to react adequately with the vulcanised structure, to promote a good heat transfer and to remove the fibres by mechanical or chemical action.

10.3 Ground rubber in civil engineering applications

Civil engineering market encompasses a wide range of uses for scrap tires, replacing some other materials currently used in construction such as lightweight fill materials, expanded polystyrene insulation blocks, drainage aggregate or even soil or clean fill [3].

The use of scrap tires in civil engineering applications is based upon the unique characteristics of tire shreds, namely lightweight, good insulation properties, very high ability to transmit water, good long-term durability and high compressibility. With these properties, engineers can use tire shreds to solve many of the construction problems and just as important, tire shreds can save their money. These applications can also consume a very large quantity of scrap tires.

Tire shreds are cost-effective substitutes for traditional materials when they are used to stabilise weak soil, such as constructing road embankments or as a subgrade fill. Additionally, tire shreds provide effective subgrade insulation for roads, walls and bridge abutments.

The other uses of scrap tire include playground surface material, gravel substitute, drainage around building foundations, building foundation insulation, erosion control/rainwater runoff barriers (whole tires), wetlands/marsh establishment (whole tires), crash barriers around race tracks (whole tires) and boat bumpers at marinas (whole tires).

References

[1]   Singleton, R., Davies, TL. Rubber technology and manufacture. Butterworth, London, UK,
       1982.
[2]   Makrov, VM., Drozdovski, VF. Reprocessing of tires and rubber wastes. Ellis Horwood, New
       York, USA, 1991.
[3]   Knorr, K. Reclaim from natural and synthetic rubber scrap for technical rubber goods.
       Kautsch Gummi Kunstst. 1994, 47, 54.
[4]   Warner, WC. Methods of devulcanization. Rubber Chem. Technol. 1994, 67, 559.

Abbreviations

ABS	Acrylonitrile–butadiene–styrene copolymer
ACS	American Chemical Society
AFM	Atomic force microscopy
aPP	Atactic polypropylene
ATR	Attenuated total reflectance spectroscopy
BA	Butyl acrylate
BMA	Butyl methacrylate
BSE	Back-scattered electrons
CCD	Charge-coupled device
CCI	Canadian Conservation Institute
DMA	Dynamic mechanical analysis
DNA	Deoxyribonucleic acid
DRIFTS	Diffuse reflectance infrared Fourier transform spectroscopy
DSC	Differential scanning calorimetry
EA	Ethyl acrylate
EMA	Ethyl methacrylate
EDS or EDX	Energy-dispersive X-ray spectrometry
EDXRF	Energy-dispersive X-ray fluorescence spectrometry
EVA	Ethylene–vinyl acetate copolymer
FTIR	Fourier transform infrared spectroscopy
GCI	Getty Conservation Institute
GC–MS	Gas chromatograpy–mass spectrometry
GPC	Gel permeation chromatography
HDPE	High-density polyethylene
HIPS	High-impact polystyrene
ICOM–CC	International Council of Museums–Committee for Conservation
IIC	International Institute for Conservation
iPP	Isotactic polypropylene
IR	Infrared (radiation)
ISO	International Organization for Standardization
IUPAC	International Union of Pure and Applied Chemistry
LDPE	Low-density polyethylene
LCST	Lower critical solution temperature
LLDPE	Linear low-density polyethylene
MA	Methyl acrylate
MMA	Methyl methacrylate
M_n	Number average molecular weight
M_w	Weight average molecular weight
NMR	Nuclear magnetic resonance
PB	Polybutadiene
PBA	Poly-n-butyl acrylate
PBMA	Polybutyl methacrylate
PE	Polyethylene
PEA	Polyethyl acrylate
PEMA	Polyethyl methacrylate
PEO	Polyethylene oxide
PET	Polyethylene terephthalate

https://doi.org/10.1515/9783110640328-011

PI	Polyisoprene
PiBA	Poly-*iso*-butyl acrylate
PiBMA	Poly-*iso*-butyl methacrylate
PIXE	Particle-induced X-ray emission
PMA	Polymethyl acrylate
PMMA	Polymethyl methacrylate
PP	Polypropylene
PPO	Polyphenylene oxide
PS	Polystyrene
PS-*b*-PMMA	Polystyrene-*b*-polymethyl methacrylate
PTFE	Polytetrafluoro ethylene (Teflon)
PVAc	Polyvinyl acetate
PVC	Polyvinyl chloride
q	Polydispersity index
RH	Relative humidity
RNA	Ribonucleic acid
SAN	Styrene–acrylonitrile copolymer
SAXS	Small angle X-ray scattering
SBR	Styrene–butadiene rubber
SBS	Styrene–butadiene–styrene copolymer
SEC	Size-exclusion chromatography
SEI	Secondary electron imaging detector
SEM	Scanning electron microscopy
SEM–EDS	Scanning electron microscopy–energy-dispersive X-ray spectrometry
SIS	Styrene–isoprene–styrene copolymer
TEM	Transmission electron microscopy
TFEMA	Trifluoroethyl methacrylate
T_g	Glass transition temperature
TGA	Thermogravimetric analysis
T_m	Melting point
TS	Tensile strength
UCST	Upper critical solution temperature
USXAS	Ultra-small angle X-ray scattering
UV	Ultraviolet (radiation)
VOC	Volatile organic compounds
WAXS	Wide angle X-ray scattering
XPS	X-ray photoelectron spectroscopy
XRF	X-ray fluorescence
YI	Yellowness index

Glossary

Abrasion resistance The resistance of a material to loss surface particles due to frictional forces applied to it. It is also the ability to resist mechanical wear.

Accelerated life test The testing of a material by subjecting it to conditions in excess of its normal service parameters in an effort to approximate the deteriorating effect of natural or in-use long-term service conditions, but in a short time.

Accelerator A chemical that speeds up vulcanisation reaction. This allows rubber to cure in a shorter time frame at a lower temperature.

Acid resistance The ability of a material to resist exposure to acids. The degree of this attack is both temperature and concentration dependant.

ACM Abbreviation for acrylic rubber.

ACN Abbreviation for acrylonitrile.

Activator A chemical used in elastomer compounding in small quantities to increase the effectiveness of an accelerator.

Adhesion The state in which two surfaces are held together by interfacial forces.

Adhesive A substance able of holding two or more materials together by surface attachment.

Ageing The irreversible change of material properties during the exposure to a deteriorating environment, including UV, oxygen, ozone, pollutants, etc. Ageing can also refer to controlled exposure of rubber samples to a variety of deteriorating factors to allow the evaluation of long-term resistance.

Ageing resistance The ability of a given material to resist deterioration of its properties caused by ageing due to oxygen, heat, light and ozone or their combinations.

Agglomerate A cluster of particles loosely held together. One of the primary roles of the mixing process is to breakdown the agglomerates, thus promoting a good dispersion.

Air traps A rubber-moulding defect due to air being trapped between the mould and the rubber or in the vulcanisate.

Amorphous A non-crystalline polymer in which the macromolecular chains are arranged in a disordered way.

Anti-degradants Materials added to a rubber compound to reduce the effects of deterioration caused by oxidation, ozone, light and/or combinations of these.

Anti-oxidant A material added to a rubber compound to reduce the deterioration caused by oxygen.

Anti-ozonant A material added to a rubber compound to reduce the damage resulting from the effects of ozone.

Anti-static agents Chemicals added to a rubber compound to dissipate the build-up of electron charges, thereby eliminating a spark or shock risk.

Anti-tack Substance applied to the surface of an elastomer to stop it from adhering either to itself or to other elastomers.

ASTM Abbreviation for American Society for Testing and Materials.

AU Abbreviation for polyester type polyurethane rubber.

Autoclave A pressure vessel that vulcanises rubber products in a pressurised steam environment.

Backrinding Tearing or distortion of a moulded rubber product at the line of separation of the mould pieces usually in the form of a raged or torn indentation.

Banbury mixer A type of internal mixer designed by F.H. Banbury.

Bank The amount of rubber adjacent to the nip of the rolls on both mills and calenders.

Barrel Part of the extruder in which the screw rotates.

Batch Product of one mixing operation in a discontinuous process.

Black Abbreviation for carbon black.

https://doi.org/10.1515/9783110640328-012

Blank A measured weight or dimension of a rubber compound suitable to fill the cavity of a compression or transfer mould. Usually the blank weight/volume is slightly higher than the finished component to allow full compression in the cavity.

Bleeding Surface exudation of a liquid or solid compounding material, often oils or lubricants, from the surface of vulcanised or unvulcanised rubber due to a partial or complete incompatibility (insolubility). It can also be referred to as leaching.

Blemish An unwanted imperfection on the surface of a moulded product.

Blister A cavity deforming the surface of vulcanisates usually due to expansion of an entrapped gas (air or other volatiles), thus determining a surface or internal imperfection.

Bloom A change in the surface appearance of an item caused by the migration of a solid (or liquid) material to the surface due to incompatibility. It normally leaves a waxy or milky deposit.

Blowing agent A compounding ingredient introduced into an elastomer which produces a gas by chemical or physical action during the processing stage. It is used in the manufacture of sponge rubber or hollow.

Bonding agent A material used to promote bonding of rubber to other materials during the processing stage.

BR Abbreviation for rubber based on butadiene.

Breakdown The first stage of the mixing process in which plasticising of raw rubber prior to the incorporation of compounding ingredients occurs.

Brittleness The tendency of an elastomer to crack when deformed or impacted.

Brittle point The lowest temperature at which a cured rubber specimen will break under a measured sudden impact and specified test conditions. This is an indication of low temperature flexibility.

Butyl rubber ASTM designation for isobutylene-isoprene rubber.

Calender A precision machine equipped with three or more heavy, internally heated or cooled rolls, parallel and revolving in opposite directions, which is used for highly accurate continuous sheeting or plying up of rubber compounds. It is often used to form rubber sheeting where thickness needs to be accurately controlled.

CAM Abbreviation for computer-aided manufacture.

Carbon black Elemental amorphous carbon in finely divided form, used to reinforce rubber compounds. The degree of reinforcement increases with a decrease in its particle size.

Catalyst A chemical, usually added to a mix in small quantities relative to the reactants, that modifies and increases the rate of a reaction.

Cavity A hollowed out area of a mould, forming the outer surface of the moulded part. Moulds may be single or multi-cavity.

Cell A single, small open space surrounded partially or completely by walls as in sponge.

Cellular rubber A generic term for rubber containing either open, closed or both types of cells dispersed throughout the material. These cells are formed by blowing agents during rubber processing.

Chalking The formation of a powdery residue on the surface of rubber which is commonly as a result of UV action onto the material surface.

CO Abbreviation for epichlorohydrin homopolymer.

Coagent An ingredient added to a rubber compound, usually in small amounts, to increase the cross-linking efficiency of non-sulphur vulcanising systems, such as organic peroxides.

Coated fabric A product constructed by coating a fabric with rubber, resulting in a flexible material which can be moulded into products or used in conjunction with rubber to provide higher rigidity and improved extrusion resistance.

Coefficient of friction Force required to move one surface over another by pressing and sliding the two surfaces together.

Coefficient of thermal expansion The fractional change in dimension of a material for a unit change in temperature.

Cold flexibility Flexibility of elastomers due to exposure to a specified low temperature for a given period of time.

Cold flow (also called Creep) Continual dimensional change that follows the initial instantaneous deformation in non-rigid materials under static load (creep). The slow deformation of raw and cured rubber may occur under gravitational force at or below room temperature, inducing a dimensional change in it.

Cold resistance The ability of an elastomer to work at low temperatures.

Compatibility The ability of different materials to blend and form a homogeneous system.

Compound A term applied to a mixture of polymers, reinforcements, fillers, curatives and other ingredients to produce a rubber material. The compound is prepared according to a prescribed formula and mixing process. Sometimes it is called **stock**.

Compression moulding A moulding process in which an uncured rubber blank is placed directly in the mould cavity and compressed to its final shape by closing the mould. This process normally results in excess material in the form of flashes.

Compression set The residual deformation in a material after the removal of a compressive stress induced for a given length of time at a specified temperature.

Conductive rubber A rubber capable of conducting electricity.

Continuous vulcanisation A process where the vulcanisation of a rubber compound takes place in a continuous manner.

Copolymer A polymer composed from two different monomers (i.e., NBR is composed by polybutadiene and acrylonitrile).

CPE ASTM designation for chlorinated polyethylene.

CR Abbreviation for chloroprene rubber.

Cracking A sharp break or fissure in the surface of rubber items due to exposure to light, heat, ozone or repeated bending or stretching.

Crazing The formation of shallow cracks on the surface of rubber, as a result of exposure to UV light or to certain chemicals. Although they look similar, crazing differs from ozone cracking, as it does not depend on the presence of an externally applied strain.

Creep (see also Cold flow) The deformation in either vulcanised or unvulcanised rubber under stress, which occurs with lapse of time after the immediate deformation. It is the time-dependent part of a strain resulting from stress.

Cross-section A section formed by a plane cutting through an object, usually at right angles to an axis.

Cross-linking (see also vulcanisation) The formation of chemical bonds among the polymer chains to give a three-dimensional network structure. Once cross-linked, the material cannot be reprocessed. It is a form of curing.

Cross-link density A measure of the relative number of cross-links in a given volume of elastomer.

Crude rubber Refers to raw rubber.

Crumb rubber Vulcanised waste or scrap rubber that has been ground down to a known mesh size and can then be added to a new compound as a filler.

Crystallinity The orientation of the disordered long-chain molecules of a polymer into repeating ordered patterns. The crystallinity degree affects stiffness, hardness, low temperature flexibility and heat resistance of any polymer.

CSM ASTM designation for chlorosulphonated polyethylene.

Curatives The collective term for the chemicals involved in curing of rubber materials, including accelerators, vulcanising chemicals such as sulphur and activators.

Curing Similar to cross-linking and vulcanisation, even if vulcanisation refers specifically to sulphur cross-linking, while cure covers all the types (sulphur, peroxide, radiation, etc.). This process results in the cross-linking of polymer chains.

Curing agent A chemical that will causes cross-linking.

Curing temperature The temperature at which curing takes place.

Curing time The required amount of time needed to complete the curing process to a pre-determined level. It is dependent on temperature, material type and section of the rubber profile.

Cycle One complete operation of a moulding machine from closing time to closing time.

Deflashing The process of removing excess material from the flash-line resulting from the moulding process. Various methods exist, including cryogenic trimming.

Degassing The passing of a gas out of a rubber, normally generated by the volatile ingredients in the rubber mix when activated at elevated temperatures.

Delamination The separation of layers of rubber (normally in a plied format) or the rubber separation from a surface to which it is bonded.

Demoulding The operation of removing a vulcanised rubber product from the mould in which it has been cured. This can be done carefully by hand, but in some cases pins or brushes can be incorporated into the mould or into the press to perform this function automatically.

Desiccant A rubber compounding ingredient used to absorb moisture irreversibly, particularly for the purpose of minimising the risk of porosity during vulcanisation.

Die A shaped plate fitted in the head of an extruder designed to create a given profile in the extrudedate.

Die swell The change in dimensions of an extruded rubber section as it exits the die. This swell is mainly due to the elastic recovery of the material.

Dimensional stability The ability of an elastomer to retain its original shape and size having been exposed to a combination of stress and temperature.

Dispersion The distribution of particles throughout a medium. For rubber, this often refers to the distribution of compounding ingredients in the rubber mix.

Dumbell A standard, flat strip specimen shaped like a dumbbell, which is used in many mechanical tests.

Durometer An instrument for measuring relative hardness of rubber. "A" durometer is used for flexible materials (i.e., rubber) and "D" for rigids, as plastics.

Dynamic properties Mechanical properties exhibited under repeated cyclic deformations.

ECO Abbreviation for epichlorohydrin copolymer with ethylene oxide.

Efficient vulcanisation A term applied to vulcanisation systems in which sulphur or a sulphur donor is used very efficiently for cross-linking the rubber.

Elastic limit The maximum extent to which a material may be deformed and yet returns to its original dimensions after removal of the deforming force.

Elasticity The rapid recovery of a material to its initial shape and size after deformation and release of the stress which caused the deformation.

Elastomer A polymeric material which, at room temperature, is capable of recovering substantially in shape and size after removal of a deforming force. This generally refers to the polymer as opposed to rubber, which preferably indicates the compounded materials.

Elongation Extension produced by tensile stress, usually expressed as a percent of the original unit length.

Elongation at break The elongation measured at the breaking point.

Embrittlement A rubber compound becomes brittle during low or high temperature exposure or as a result of ageing.

Endothermic A chemical reaction that absorbs heat energy.

EPDM Abbreviation for ethylene-propylene-diene rubber.

Exothermic A chemical reaction in which heat energy is liberated.

Extender An inert material added to a rubber compound to increase the volume and to lower the cost of the compound without imparting any enhanced physical properties.

Extraction The process of removing one or more components of a homogeneous mixture by treating it with a solvent in which the components to be removed are soluble but not the mixture as a whole.

Extrudate The profiled material that results from the extrusion process.

Extruder A machine designed with a driven screw to create a continuous profiled rubber shape by forcing the rubber through a die which has a shape similar to that of the required profile.

Exudation Delayed phase separation of incompatible material from rubber, also called bleeding, blooming, spewing or sweating.

Fatigue The weakening of an elastomer during repeated deformation, strain or compression.

FEA Abbreviation for finite element analysis.

FEPM Abbreviation for tetrafluoroethylene/propylene copolymers.

FFKM Abbreviation for perfluoroelastomers.

Filler A compounding ingredient which is added to a rubber usually in finely divided form. There are two main categories of filler: reinforcing, which adds strength to the elastomer (see **Reinforcing fillers**) and extending (non-reinforcing), which has the function of cheapening the elastomer (see **Extender**).

Finite element analysis A mathematical technique developed to predict the stress–strain behaviour of objects.

Fire retardant An additive used in rubber compounding to reduce fire hazard.

FKM Abbreviation for fluorocarbon rubber.

Flame resistance A material's resistance to burning.

Flash The excess material resulting from the moulding operation found at the mould split lines.

Flex cracking A cracking condition of the surface of vulcanisates such as tires and footwear, resulting from constantly repeated bending or flexing of the part.

Flex life The number of cycles required to produce a specified state of failure in a rubber specimen.

Flexural strength Ability of an elastomer to flex without permanent distortion or damage.

Flow marks Marks present on the surface of vulcanisates caused by insufficient or improper flow of the material in the moulding cavity.

Foam rubber Cellular rubber formed by whipping latex to a froth and then vulcanised.

Formulation Kinds and proportions of ingredients for a mix, together with the method of incorporation. It is also called recipe.

Free sulphur Portion of sulphur originally present in the compound, which did not react during vulcanisation.

Gate The point through which a rubber is injected into a moulding cavity in both transfer and injection moulding techniques.

Gate mark A witness mark left on the moulding as a result of injecting rubber through the gate. This can be either a raised or sunken mark on the surface of the moulding.

Glass transition temperature (T_g) The point at which any polymer loses its flexibility at low temperature.

Green stock Raw unvulcanised compound.

Green strength The strength of rubber in uncured state.

Hardness Measurement of resistance to indentation. The most common units are Shore A and Shore D.

Heat ageing Accelerated ageing in air or oxygen at elevated temperature and, in some cases, pressure for specified periods of time. The deterioration is generally noted as a percent change from the originally measured properties.

Heat build-up The generation of heat due to hysteresis when rubber is rapidly or continually deformed.

Heat embrittlement The hardening of a vulcanised rubber compound when aged at elevated temperatures in air, accompanied by increase in modulus and decrease in tensile and elongation.

Heat history The total heat received by a rubber compound during its transformation (i.e., mixing, milling, extruding, calendering). Particularly, it refers to the temperatures reached by the rubber compound and the time it has been held at these temperatures. Since vulcanisation takes place at elevated temperatures, incipient cure or scorch can take place if heat history is excessive.

Heat resistance The ability of rubber to undergo elevated temperature without varying its original properties.

HNBR Abbreviation for hydrogenated nitrile rubber.

Homopolymer A polymer formed from a single monomer.

Hooke's law Within the limits of elasticity of a material, tension is proportional to elongation, or strain is proportional to the stress producing it.

Hysteresis The heat generated by rapid deformation of a vulcanised rubber part. It is the difference between the energy of the deforming stress and that of the recovery cycle. The loss of energy results in heat build-up.

Hysteresis loss The loss of mechanical energy due to hysteresis.

IIR ASTM designation for isobutene-isoprene rubber or butyl rubber.

Impact resistance The material resistance to fracture under a quickly applied load.

Impact strength A measure of the toughness of a material to rapidly applied loads. It is often represented as the energy required to break a specimen with a single swinging blow.

Incompatibility Inability of a material to form a homogeneous system.

Inert A material that has little tendency to reinforce or have any other effect upon the properties of vulcanisates.

Inhibitor A compounding ingredient that is added to a mix to slow down or prevent rubber curing.

Injection moulding The moulding operation wherein a rubber compound is preheated in the barrel of an extruder and injected under pressure through a series of runners into the mould cavity while it is in a plastic state.

Insert Normally a metal or plastic component to which rubber is chemically and/or physically bonded during the moulding process for a definite purpose.

Internal mixer An enclosed mixing machine for rubber inside of which there are two specially shaped counter-rotating heavy-mixing rotors with small clearance between themselves and the enclosing walls. The mixing chamber is jacketed and may be heated or cooled, as may be the rotor. Rotors masticate rubber and incorporate compounding materials through the action of mechanical work (shear) with the aim of creating a homogenised finished rubber ready for the vulcanisation process.

ISO Abbreviation for International Organisation for Standardisation.

Knit lines, weld lines or marks Imperfections in a vulcanised item where material did not flow properly during moulding. Also called poor knitting.

Knots Lumps that appear in a stretched rubber part, generally due to poor dispersion of a curative, such as sulphur.

Latex It refers to the colloidal emulsion obtained from *Hevea* tree.

Leakage rate The rate at which a fluid passes through or around a seal.

Light ageing Deterioration of any material when exposed to light (direct or indirect, natural or artificial).

Liquid silicone rubber High-purity platinum-cured silicone with low compression set, great stability and ability to withstand extreme heat and cold.

Litharge Lead monoxide, PbO, formerly used as an inorganic accelerator but now mainly used as a vulcanising agent in some polychloroprene rubber.

Loading The kind and quantity of fillers mixed with raw rubber.

Low temperature flexibility The ability of an elastomeric product to be flexed at low temperatures without cracking.

Lubricants Chemicals mixed into a compound (internal) to reduce the tendency to stick to the processing equipment or to lower the heat build-up on flexing.

Masterbatch A homogeneous mixture of rubber and one or more materials in high proportions for use as a raw material in the final mixing of a compound. Masterbatches are used to improve the dispersion of reinforcing agents, rubber breakdown, to lower the heat history of a compound and to facilitate the weighing and dispersion of small amounts of additives.

Mastication The breakdown or softening of raw rubber by the mechanical work (shear) done in a mill mix or in an internal mixer. This can be accelerated by the use of a peptiser.

Memory Ability of rubber to return to its original shape after deformation.

Mill mixer A machine used for the mechanical mixing or working (and sheeting) of rubber consisting of two adjacent, heavy, hardened-steel, counter-rotating horizontal rolls. Generally, they revolve at slightly different speeds.

Mineral oils Petroleum and other hydrocarbon oils obtained from mineral sources, which in rubber compounds act as softeners and extenders.

Modulus In elastomer technology, this is defined as the stress at a particular strain or elongation. Modulus tends to increase with hardness. Higher modulus materials are more resistant to deformation and extrusion.

Modulus of elasticity The ratio of stress to strain in an elastic material. It is also called Young's Modulus.

Mould cavity Profiled shape cut into a mould within which rubber is cured to produce an item.

Mould marks An imperfection transferred to a vulcanisate from corresponding marks present on a mould surface.

Mould release A substance applied to the surface of a mould cavity to aid the release of vulcanisate since they reduce the tendency to stick to a mould.

Mould shrinkage Dimensional loss in vulcanisate that occurs during cooling after it is removed from the mould.

NBR Abbreviation for nitrile-butadiene rubber.

Necking The localized reduction in cross-section that may occur in a material under tensile stress.

Neoprene Originally the trade name, now the generic name, of polymers and copolymers based on chloroprene.

Nerve The elastic resistance and toughness of raw rubber or compounds to permanent deformation induced during processing.

Nip The radial distance between the rolls on a mill or calender, measured at the line of centres.

Non-staining An accelerator, antioxidant or similar substance that will not discolour other goods placed next to the rubber in which it is used.

NR Abbreviation for natural rubber.

Open cell A condition of cellular or sponge rubber where the cell walls are not continuous but interconnected, rendering the sponge breathable and able to soak up fluids.

Open steam cure A method of vulcanisation in which the steam comes in direct contact with the product being vulcanised, generally in an autoclave.

Optimum cure The state of vulcanisation (time and temperature) at which a desired property value or combination of property values is obtained. In some rubber, this may require post-curing or autoclaving to produce the desired level of cure.

Orange peel The description of the surface appearance of a cured part of rubber when stretched, resembling the surface of an orange. Generally, it is the result of a poor dispersion of the filler or of the incomplete breakdown of rubber.

Outgassing The release of vapours or gases from a rubber compound.

Over-cure A degree of cure greater than the optimum, resulting in worsening of certain physical properties. In some cases, this can lead to a loss of elongation and an increase in hardness. In the case of natural rubber this can lead to reversion.

Oxidation The reaction of oxygen with a material, usually accompanied by a change in feel, appearance of surface or change, usually adverse, in physical properties.

Ozone Gaseous allotropic form of oxygen having a characteristic odour, which is a powerful oxidising agent. It is present in the atmosphere at low levels and causes cracking in certain types of elastomeric compounds.

Ozone resistance The ability to withstand the deteriorating effect of ozone.

Parting line The line on the surface of a moulded part where the separated mould parts meet and create a small clearance gap.

PB Abbreviation for polybutadiene.

Peptiser A compounding material used in small proportions to accelerate, by chemical action, the breakdown and softening of rubber under the influence of mechanical action or heating (or both).

Permeability Measure of the ease with which a liquid or gas can pass through a rubber material.

Peroxide A very reactive compound containing a bivalent −O−O− group in the molecule. It is used as cross-linking agent for some rubber.

Phase A physically homogeneous, mechanically separable portion of a material system.

Phr Abbreviation for parts per hundred of rubber, used for indicating the proportions of ingredients in a rubber formulation.

Pigment The material used to impart colour.

Plastic deformation Deformation of a plastic material beyond the point of recovery, accompanied by continuing deformation with no further increase in stress.

Plasticiser A substance, usually a heavy liquid or oil, which is added to an elastomer to decrease stiffness, improve low temperature properties, reduce cost and/or improve processing.

Plasticity The tendency of a material to remain deformed after reduction of deforming stress to or below its yield stress.

Plateau effect The maintenance of rubber properties over a broad range of vulcanisation times.

Platen The flat steel or cast iron part of a moulding machine that applies heat and pressure to a mould.

Ply A layer in a laminated structure.

Ply adhesion The force required to separate two adjoining plies.

Poisson's ratio The measure of the simultaneous change in elongation and in cross-sectional area within the elastic range during a tensile or compressive test.

Polymer A macromolecular material formed by the chemical combination of monomers having either the same or different chemical composition.

Polymerisation A chemical reaction in which the monomers are linked together to form long chains whose molecular mass is a multiple of the original substance. When two or more different monomers are involved, the process is called copolymerisation. Polymerisation may be via emulsion (SBR, nitrile), ionic (butyl), solution (EPDM, cis-polyisoprene), suspension or bulk (PVC) processes.

Post-cure Heat or radiation treatment, or both, to which a cured or partially cured thermosetting plastic or rubber is subjected to increase the state of cure or to enhance the level of one or more properties.

Pre-curing Premature vulcanisation taking place prior to vulcanisation. It is similar to scorch.

Pre-form The pre-shaped, uncured material placed in a mould for vulcanisation. Pre-shaping is done to facilitate mould flow, control weight accuracy and so on.

Primer A coating applied to the surface of a material prior to the application of an adhesive to improve the performance of the bond.

Processability The relative ease with which raw or compounded rubber can be processed. This can relate to all aspects of manufacturing.

Processing aid A compounding ingredient that is added to a mix with the aim of improving a material's ability to be processed.

PU Abbreviation for polyurethane.

Rate of cure The relative time required to reach a predetermined state of vulcanisation under specified conditions.

Raw rubber Unprocessed, vulcanisable elastomer.

Reinforcement The stiffening effect of chemicals, such as carbon black, on unvulcanised elastomer mixtures and the enhancement of the physical properties of the vulcanised compound, such as tensile, elongation, modulus, abrasion resistance, tear and so on.

Reinforcing filler A compounding ingredient added to rubber to increase its mechanical resistance.

Resilience The ratio between the energy output and energy input in a rapid full recovery of a deformed rubber specimen.

Retarder A compounding ingredient that is added to a mix to reduce the tendency of a rubber compound to vulcanise prematurely. It is also called scorch **inhibitor**.

Reversion Deterioration of physical properties that may occur upon excessive vulcanisation of some elastomers, highlighted by a drop in hardness and tensile strength and by an increase in elongation. Natural rubber, butyl, polysulphides and epichlorohydrin polymers exhibit this effect.

Rheometer Instrument for the study of vulcanisation behaviour of rubber.

RTV Abbreviation for room temperature vulcanisation.

Saturation Saturated chemical compounds are those whose constituent molecules contain no double or triple valency bonds.

SBR Abbreviation for styrene-butadiene rubber.

Scission Breaking of chemical bonds.

Scorching Premature vulcanisation of a rubber compound generally due to an excessive heat history.

Screw Rotating member with a helical groove to propel rubber through the barrel of an extruder.

Secondary accelerator An accelerator used in smaller concentrations, when compared to the primary accelerator, to achieve a faster rate of vulcanisation.

Shelf life An expression describing the time for which an unvulcanised stock can be stored without losing any of its processing, curing or physical properties.

Shrinkage The reduction in size upon cooling of a moulded rubber part.

Sink mark A depression in the surface of a vulcanised item caused by the collapse of blister or bubble or by internal shrinkage in vulcanisates.

Sponge rubber Rubber showing a porous structure, with cells being open and intercommunicating.

Sprue The primary feed channel that runs from the outer face of an injection or transfer mould to the mould gate in a single cavity mould or to runners in a multiple cavity mould. It also corresponds to the piece of material formed or partially cured in the primary feed channel that remains attached to the vulcanised item before finishing.

Sprue mark A mark in relief left on the surface of an injection or transfer moulded part, after removal of the sprue.

Stabiliser A chemical used to prevent or retard rubber degradation by heat, light or atmospheric exposure.

Staining Change of rubber colour when exposed to light or change of colour of a material in contact with, or adjacent to, rubber.

State of cure The degree of vulcanisation of a rubber compound.

Steam cure A method of vulcanising rubber parts by exposing them directly to steam.

Stock A term for unvulcanised rubber compound.

Strain Deformation resulting from a stress.

Stress Force per cross-sectional area that is applied to a specimen.

Stress relaxation The time-dependent decrease in stress for a specimen under a constant strain.

Substrate A material upon the surface of which an adhesive is spread for any purpose, such as bonding or coating.

Sulphur Vulcanising agent responsible for the cross-linking of many rubber.

Tack The property that causes contacting surfaces of unvulcanised rubber to stick to each other.

Tackifier A compounding material that enhances the ability of vulcanised rubber to adhere to itself or to another material.

Tear resistance Resistance to the growth of a nick or cut in a rubber specimen when tension is applied.

Tear strength The maximum force required to tear a test specimen.

Tensile strength A measure of the stress required to break a standard test specimen.

Thermal degradation An irreversible change in the properties of a material due to exposure to heat.

Thermal expansion Linear or volumetric expansion of a material caused by increase in temperature.

Thermoplastic Polymer that melts due to the application of heat, thus allowing its reshaping.

Thermoplastic elastomer (TPE) A diverse family of rubber-like materials that, unlike conventional vulcanised rubber, can be processed and recycled like thermoplastic materials since they contain physical cross-links.

Thermoset Materials that undergo chemical cross-linking of their molecules when processed, and cannot be softened and reshaped following further application of heat. They are insoluble and infusible.

Tolerance The amount by which a property of a material or object can vary from a specified value and still be acceptable.

Transfer moulding The process of moulding by forcing rubber from a reservoir-heated chamber through a gate into the moulding cavity of a closed mould.

Trapped air The air that is enclosed in a product or between a mould surface and an item during vulcanisation.

Tumbling A finishing process for removing flash from a moulded part by placing it in a rotating barrel with or without the added finishing material such as shells, dry ice and so on.

Under-cure A condition where rubber has not been cured to its optimum state and thus exhibits a reduction in its physical properties.

UV absorber A compounding material that retards the deterioration caused by sunlight and other UV light sources.

Viscoelasticity A combination of viscous and elastic properties in a polymer.

Viscosity The resistance of a material to flow due to either gravity or stress.

Vulcanisate Preferably used to denote the product of vulcanisation, without reference to shape or form.

Vulcanisation An irreversible process during which a rubber compound, through a change in its chemical structure (cross-linking), becomes less plastic and more elastic and more resistant to swelling by organic liquids. Elastic properties are conserved, improved or extended over a greater range of temperature. It often refers to the reaction of rubber specifically with sulphur, while curing covers other methods of cross-linking. Both terms are often used interchangeably.

Vulcanising agent Any material that can cause in rubber a change in physical properties know as vulcanisation, such as sulphur, peroxides, metal oxides and so on.

Weathering The tendency of rubber to surface crack under exposure to atmospheres containing ozone and other pollutants.

Index

https://doi.org/10.1515/9783110640328-013

www.ingramcontent.com/pod-product-compliance
Lightning Source LLC
Chambersburg PA
CBHW061420210326
41598CB00035B/6275